Value-Based Management with Corporate Social Responsibility

Second Edition

FINANCIAL MANAGEMENT ASSOCIATION
Survey and Synthesis Series

The Search for Value: Measuring the Company's Cost of Capital
Michael C. Ehrhardt

Managing Pension Plans: A Comprehensive Guide to Improving Plan Performance
Dennis E. Logue and Jack S. Radar

Efficient Asset Management: A Practical Guide to Stock Portfolio Optimization and Asset Allocation
Richard O. Michaud

Real Options: Managing Strategic Investment in an Uncertain World
Martha Amram and Nalin Kulatilaka

Beyond Greed and Fear: Understanding Behavioral Finance and the Psychology of Investing
Hersh Shefrin

Dividend Policy: Its Impact on Firm Value
Ronald C. Lease, Kose John, Avner Kalay, Uri Loewenstein, and Oded H. Sarig

Value Based Management: The Corporate Response to the Shareholder Revolution
John D. Martin and J. William Petty

Debt Management: A Practitioner's Guide
John D. Finnerty and Douglas R. Emery

Real Estate Investment Trusts: Structure, Performance, and Investment Opportunities
Su Han Chan, John Erickson, and Ko Wang

Trading and Exchanges: Market Microstructure for Practitioners
Larry Harris

Valuing the Closely Held Firm
Michael S. Long and Thomas A. Bryant

Last Rights: Liquidating a Company
Ben S. Branch, Hugh M. Ray, Robin Russell

Efficient Asset Management: A Practical Guide to Stock Portfolio Optimization and Asset Allocation, Second Edition
Richard O. Michaud and Robert O. Michaud

Real Options in Theory and Practice
Graeme Guthrie

Value-Based Management with Corporate Social Responsibility, Second Edition
John D. Martin, J. William Petty, and James S. Wallace

Value-Based Management with Corporate Social Responsibility

Second Edition

John D. Martin

J. William Petty

James S. Wallace

2009

OXFORD
UNIVERSITY PRESS

Oxford University Press, Inc., publishes works that further
Oxford University's objective of excellence
in research, scholarship, and education.

Oxford New York
Auckland Cape Town Dar es Salaam Hong Kong Karachi
Kuala Lumpur Madrid Melbourne Mexico City Nairobi
New Delhi Shanghai Taipei Toronto

With offices in
Argentina Austria Brazil Chile Czech Republic France Greece
Guatemala Hungary Italy Japan Poland Portugal Singapore
South Korea Switzerland Thailand Turkey Ukraine Vietnam

Published by Oxford University Press, Inc.
198 Madison Avenue, New York, New York 10016

www.oup.com

Library of Congress Cataloging-in-Publication Data
Martin, John D., 1945–
Value-based management with corporate social responsibility / John D. Martin,
J. William Petty, and James S. Wallace. — 2nd ed.
p. cm. — (Financial management association survey and synthesis series)
Includes bibliographical references and index.
ISBN 978-0-19-534038-9
1. Value analysis (Cost control) 2. Industrial management.
I. Petty, J. William, 1942– II. Wallace, James S. III. Title.
HD47.3.M37 2009
658—dc22 2008049916

9 8 7 6 5 4 3 2 1
Printed in the United States of America
on acid-free paper

Preface

The thesis of this book is simple. Value-based management (VBM), done from the perspective of a firm's owners, is socially responsible. Moreover, "getting it right" requires that managers "keep score" by using a numerical measure of value creation. However, the measure is a reflection of the firm's success or failure in achieving value creation, not a guide to action. Much of the first edition of this book delved into the nuances of getting the measure right, which, at the time, was the center of the VBM debate. The hope was that the VBM holy grail would be found in a more sophisticated measure of VBM success, a better scorecard.

In the eight years since the publication of the first edition of this book we have learned a lot about VBM that has led us to write a revised edition:

- First, VBM's focus on maximizing shareholder wealth is as relevant today as it has always been. Value-based management is the visible representation of Adam Smith's invisible hand that directs resource allocation in a capitalistic economy. However, VBM is very hard to implement and requires much more than the adoption of an incentive bonus system that uses a mechanical performance measure like EVA—or any other measure for that matter.
- Second, proponents of the stakeholder theory of the firm and its modern-day incarnation as corporate social responsibility (CSR)

have much to offer VBM. What CSR brings to VBM is the simple but powerful notion that, for any firm to be successful in creating wealth, all of those who have a stake in the business must receive their share of the economic pie. Furthermore, by using an adaptation of the balanced scorecard toolkit, the needs of all these groups can be recognized and incorporated into the firm's managerial processes for measuring and evaluating its relationships with all important constituencies. Value-based management simply requires the firm to have the maximization of its long-term market value as its overriding goal.

- With regard to the second point, we have shifted our emphasis from the firm's goal of maximizing shareholder value to maximizing long-term market value. This is a not-so-subtle shift. We contend that a win-win situation can be accomplished by making the pie bigger for all of the firm's stakeholders, whether they be shareholders, customers, employees, suppliers, communities, or society in general. Since the shareholders are the ultimate owners of the firm, they will certainly benefit if the firm's value is increased, and other stakeholders will also potentially benefit: a true win-win situation.

Many events have affected the practice of management in the last decade, but probably no set of events has had a larger impact than the series of scandals involving firms such as Enron and WorldCom. In the aftermath of these scandals we have witnessed a general resistance to outwardly portraying the corporate mission as having a primary focus on shareholder wealth creation. The set of principles that serve as the foundation of VBM and its focus on shareholder value became linked to the corruption that pervaded the top management suites of these scandal-plagued firms. In the wake of these scandals the new mantra of corporate social responsibility appears to be replacing that of shareholder wealth creation as if they are somehow substitutes. The premise of this book is that VBM and CSR are in fact complements rather than substitutes. Properly done, VBM is the socially responsible management of a firm from the perspective of stockholders and society as a whole.

Therefore, unlike the first edition of *Value-Based Management,* which simply accepted the premise that firms should be managed to enhance shareholder value as a given, in part I of this book we explore fundamental questions about the modern for-profit corporation: What is its purpose, and what is the proper objective for managing the for-profit enterprise? There is general agreement that these corporations exist because it is far more efficient for them to deliver goods and services than it would be for individuals to contract with one another for these same enterprises. Nonetheless, the questions of what a corporation's goal should be and how it should be measured meet with far less agreement. The preeminent view (perhaps best associated with Nobel laureate economist Milton Friedman) is that corporations should be judged by how well they deliver shareholder value, the VBM concept. The shareholders, after all, are the owners of the corporation. This view may be widely held, but it is certainly not universally accepted. Some feel that the corporation is dependent on many stakeholders and that shareholders are only one stakeholder group. Other such groups include customers, employees, suppliers, communities, governments, the environment, and society in general. Followers of this stakeholder theory believe the needs of all of these groups must be balanced and that simply maximizing shareholders' interests is insufficient.

While we are not at all surprised that VBM is not universally accepted, we are quite surprised at the level of confusion among its advocates. This misunderstanding ranges from how to deliver shareholder value to how to measure it.

We have written this book in order to clarify this uncertainty. This book has two primary origins. The first is the authors' collective teaching and research, which applies the fundamental principles of business valuation to the management of the business enterprise for almost a half century. The book was also inspired by the results of a study carried out by the American Productivity and Quality Center's (APQC) International Benchmarking Clearinghouse in 1996, in which two of this book's authors were coresearchers.[1] That study was a large-scale research effort designed to document the practices of a broad sample of firms that had successfully implemented a VBM system. Thus, the objectives of this book are twofold: First, we synthesize a VBM model, and, second, we report the lessons learned by a number of companies that have implemented VBM programs.

In part I of this book we explain why shareholder value maximization, the premise underlying VBM, represents the goal that public corporations should strive to attain. In doing so we argue that shareholder value maximization does not simply benefit the shareholders (at the expense of other stakeholders) but society in general as well. In this discussion we also present our viewpoint of where stakeholder theory and the related concept of corporate social responsibility fit into the VBM paradigm.

In this first section we introduce VBM in general terms and pay specific attention to measurement. While there is no one perfect way to measure shareholder value, it is safe to say that there are many metrics in use that simply fail to adequately provide the proper information and incentives for use in a VBM program. We have all heard the old adage "you get what you measure and reward." Although it is doubtful that properly measuring and rewarding shareholder value is sufficient for success, we believe that properly measuring (and rewarding) shareholder value is a necessary condition for success.

In part II we look at VBM metrics in more detail. First we consider free cash flow, the underlying concept behind all of the VBM metrics. We then consider what is arguably the most popular of the VBM metrics today, economic value added (EVA). Finally, we also discuss the concept of corporate social responsibility in more detail. We find that VBM and CSR are intertwined and that CSR acts as a guide to the drivers of the VBM program, our term for which is *value(s)-based management* (V_SBM).

Part III explores two primary applications of a VBM metric: project evaluation and incentive compensation. We show that VBM metrics provide a consistent measurement tool for both project selection and project evaluation. This is a significant improvement over traditional methods, which use one measure (typically discounted cash flows) for project selection and others such as return on invested capital or operating earnings for follow-up evaluation. The use of a single VBM metric for both project evaluation and incentive compensation is the cornerstone for an entire VBM program.

A proper metric to measure shareholder value is certainly necessary to judge success; however, it provides only a score. In part IV of this book we consider the process of scoring. We believe this area is the most misunderstood within the whole VBM debate and the single biggest reason that many shareholder value implementations fail. Here we present our view, which has developed over the years as a result of conducting numerous studies and working with many firms. Our explanation of the success of certain firms and the failure of others expands upon the traditional reasons widely held and presented by shareholder value consultants.

We support the theory presented in this book with evidence from the body of academic research on VBM, and we also feel compelled to provide case study examples in addition to academic theory. Our focus is on identifying "lessons learned" so that the potential adopter might benefit from the experiences of others. We wish to avoid the common trap of believing that theory and practice should be the same because, in practice, they often are not.

Contents

Value-Based Management with Corporate Social Responsibility

Second Edition

Part I

Value-Based Management, Corporate Social Responsibility, and the Purpose of the Corporation

Why should a corporation exist? What is its purpose? Answering these fundamental questions is the objective of the first part of this book. We explore such diverse ideas as "the corporation exists only to make money" and "the corporation should make money only so that it can do something bigger or better." Answers to these questions are critical to our further study of value-based management (VBM) and corporate social responsibility (CSR), the combination of which we refer to as *value(s)-based management* (V_SBM). In addition to introducing the concepts of VBM and CSR, we also discuss the critical need for measurement in business. We argue that you cannot manage what you cannot measure and that you get what you measure and reward.

Chapter 1

The Purpose of a Corporation

> An important task for top management in the next society's corporation will be to balance the three dimensions of the corporation: as an economic organization, as a human organization, and as an increasingly important social organization.
>
> —Peter F. Drucker, *The Daily Drucker*

We begin our look into value(s)-based management (V_sBM) by first considering the fundamental question of what goal public for-profit corporations should set for themselves.[1] This is not a trivial question, and it deserves careful thought. Should corporations strive to maximize value for their shareholders? Or is such a goal synonymous with greed? Would a more appropriate objective be the betterment of society? If so, which stakeholders, employees, customers, communities, governments, or other entities should take priority? If the answer to this question is none, how should tradeoffs among the conflicting needs of these various groups be made? Or is maximizing value for shareholders somehow the same thing as enhancing society in general? These are not easy questions, but answers to them are essential to determining whether and how to implement value(s)-based management.

Adam Smith and the Invisible Hand

One stream of thought on these questions goes back to 1776, when Adam Smith published his seminal work titled the *Wealth of Nations*:

> Every individual endeavors to employ his capital so that its produce may be of greatest value. He generally neither intends to promote the public interest, nor knows how much he is promoting it. He intends only his own security, only his own gain. And he is in this led by an invisible hand to promote an end, which has no part of his intention. By pursuing his own interest he frequently promotes that of society more effectually than when he really intends to promote it.
>
> Adam Smith, *The Wealth of Nations* (1776/1999, 32)

Adam Smith used the invisible hand metaphor to illustrate that individuals (or corporations) seek wealth by following their own self-interests and in so doing create wealth for the economy as a secondary effect. This in turn benefits society as a whole. How is this accomplished? According to Smith's theory, as long as consumers are allowed to freely choose what to buy and sellers are allowed to freely decide what to sell and how to produce it, the market will determine the products sold and the prices charged on the basis of what is beneficial to the entire community. This naturally follows from the fact that all of these participants follow their own self-interest. Firms will adopt efficient methods of production in order to maximize profits. They will also be forced to restrain prices because of competitor pricing. Most important to our discussion, investors will invest in firms that are the most successful in providing returns and withdraw capital from those that are less efficient in creating value. Capital will therefore flow to those firms that are most able to create the greatest wealth for the economy, and all of this will take place automatically as if guided by an invisible hand.

The idea that profit maximization produces the best outcome for society in general can be shown by a simple example.[2] For instance, a firm uses resources from the economy in the form of materials, labor, and capital and purchases these inputs from their owners through voluntary exchanges. The firm then combines these inputs to produce goods and services that it sells to customers, again through voluntary exchanges. Since all exchanges are voluntary, we can conclude that the buyer and seller each place a value on the exchanged item that is equal to the price paid. Therefore, if the firm is able to sell the goods and services at a price that is higher than what it cost the firm to produce the goods and services (i.e., at a profit), value has been added to the economy. The more profit the firm produces, the more value is added to society.

Whether one wishes to label this self-interested behavior as greed is not really important. A short passage from the 1987 movie *Wall Street* introduced us to the concept that greed can be good:

> The point is, ladies and gentlemen, that greed, for lack of a better word, is good. Greed is right; greed works. Greed clarifies, cuts through, and captures the essence of the evolutionary spirit. Greed, in all of its forms, greed for life, for money, for love, knowledge—has marked the upward surge of mankind, and greed, you mark my words—will save not only Teldar Paper but that other malfunctioning corporation called the USA.
>
> Gordon Gekko (played by Michael Douglas) in *Wall Street* (1987)

One unfortunate outcome in the post-Enron environment is the perception that value-based management is somehow synonymous with greed, which itself is synonymous with taking advantage of other stakeholders. We believe this perception is the result of a misunderstanding of VBM. Both Gordon Gekko and Adam Smith believed that when people are free to pursue their own interest, they will fare better than they will under a system that dictates what is "good." In the process, inefficiencies are eliminated, and capital is allocated to those areas where it will most benefit the greater society.

One of the most extreme views related to the invisible hand concept suggests that the only valid purpose of a corporation is to maximize shareholder value and that anything else is irresponsible. This belief can best be represented by the following quote by Nobel laureate Milton Friedman: "Few trends could so thoroughly undermine the very foundation of our free society as the acceptance by corporate officials of a social responsibility other than to make as much money for their stockholders as possible" (Milton Friedman, *Capitalism and Freedom* [1962]).

A Stakeholder Perspective

Friedman's view is not universally accepted. Dave Packard, cofounder of the Hewlett Packard Company, has suggested a different reason for a company's existence:

> I think many people assume, wrongly, that a company exists simply to make money. While this is an important result of a company's existence, we have to go deeper and find the real reason for our being. As we investigate this, we inevitably come to the conclusion that a group of people get together and exist as an institution that we call a company so that they are able to accomplish something collectively that they could not accomplish separately—they make a contribution to society, a phrase which sounds trite but is fundamental.
>
> Dave Packard, Hewlett Packard Company, quoted in Collins and Porras (2002, 56)

Packard's view was broadened and later articulated by Edward Freeman, who in his 1984 book *Strategic Management: A Stakeholder Perspective* developed what has come to be known as *stakeholder theory.* Stakeholder theory addresses the question of what or who really matters to corporations. In contrast to Friedman's shareholder value theory, which posits that corporations should be concerned only about their shareholders, stakeholder theory posits that corporations have multiple stakeholders, all of which are important and need to be rewarded. Stakeholders include not only the shareholders but also customers, employees, suppliers, the community, the government, the environment, and society in general.

Stakeholder theory is closely linked with the Corporate Social Responsibility (CSR) movement, another concept that has gained significant attention in recent years. Like stakeholder theory, CSR implies that an organization identifies its various stakeholder groups and then attempts to balance their respective needs within the organization's overall strategy. Many of Friedman's followers advocate doing for nonshareholder stakeholders only what is required by regulation and law. In Friedman's words this means "staying within the rules of the game." In contrast, CSR advocates believe that considering the needs and values of all stakeholder groups offers strategic advantages.

Value(s)-Based Management: A Middle Ground

While Friedman's and Freeman's views potentially represent opposite ends of a continuum, others have advocated a middle ground. Perhaps the greatest mind to address these issues belonged to the father of modern management, Peter F. Drucker, who defined the purpose of a business in terms of the customer:

> If we want to know what business is, we have to start with its purpose. And the purpose must lie outside the business itself. In fact, it must lie in society, since a business enterprise is an organ of society. There is only one valid definition of business purpose: to create a customer. The customer is the foundation of a business and keeps it in existence. He alone gives employment. And it is to supply the customer that society entrusts wealth-producing resources to the business enterprise.
>
> Peter Drucker, *The Practice of Management* (1954, 37)

Although Drucker weighed in on social responsibility and incorporated the views of both Friedman and Freeman, in 2002 he brought these divergent views together in his book *Managing in a Time of Great Change*:

> A business that does not show a profit at least equal to its cost of capital is socially irresponsible; it wastes society's resources. Economic profit performance is the base without which business cannot discharge any other responsibilities, cannot be a good employer, a good citizen, a good neighbor. But economic performance is not the only responsibility of a business.... Every organization must assume responsibility for its impact on employees, the environment, customers, and whomever and whatever it touches. That is social responsibility. But we know that society will increasingly look to major organizations, for-profit and nonprofit alike, to tackle major social ills. And that is where we had better be watchful, because good intentions are not always socially responsible. It is irresponsible for an organization to accept—let alone pursue—responsibilities that would impede its capacity to perform its main task and mission or to act where it has no competence.
>
> Peter Drucker, *Managing in a Time of Great Change* (2002b, 84)

Drucker went on to state in *A Functioning Society*:

> We no longer need to theorize about how to define performance and results in the large enterprise. We have successful examples.... They do not "balance" anything. They maximize. But they do not attempt to maximize shareholder value or the short-term

> interest of any one of the enterprise's "stakeholders." Rather, they maximize the wealth-producing capacity of the enterprise. It is this objective that integrates the short-term and long-term results and that ties the operational dimensions of business performance—market standing, innovation, productivity, and people and their development—with the financial needs and financial results. It is also this objective on which all the constituencies—whether shareholders, customers, or employees—depend for the satisfaction of their expectations and objectives.
>
> Peter Drucker, *A Functioning Society* (2002a, 134)

The views of Peter Drucker are quite similar to those voiced by Michael Jensen, who defined *enlightened value maximization.* Jensen argues that, in the absence of externalities and monopoly, two hundred years of work in economics and finance have demonstrated that social welfare is maximized when each firm maximizes its market value. He argues that, unfortunately, while value maximization should be a corporation's objective function, it does not provide management with a strategy on how to do so. It is merely a scorecard. Jensen further argues for a principle of enlightened value maximization since "we cannot maximize the long-term market value of an organization if we ignore or mistreat any important constituency" (2001, 16). In other words, enlightened value maximization "utilizes much of the structure of stakeholder theory but accepts maximization of the long run value of the firm as the criterion for making the requisite tradeoffs among its stakeholders" (9). Jensen's enlightened value maximization is analogous to our value(s)-based management.

Summary: Creating Firm Value—Think Value(s)-Based Management

Some would argue that a firm exists only to make money, whereas others would counter that it should make money only so that it can do something bigger or better. The former group sees making money as the goal, while the latter sees money as only a means to an end. Both would argue, vehemently at times, that it is an either-or proposition: "If I am right, then you must be wrong."

Value-based management, as practiced over the past two decades, is about creating economic or intrinsic firm value. Its opponents suggest that it is about greed. While we do not deny the propensity of some humans to want more at the expense of others, greed is not a necessary or even a desirable quality of VBM.

Corporate social responsibility, a growing movement in recent years, has most often been presented as a firm's "responsibility" to do good or to give back to society. It just feels right. However, we believe that CSR is not primarily about an obligation but, more important, provides a process for structuring win-win agreements by the different constituencies (stakeholders) in sharing the value. We believe it is not about being altruistic. It is what works.

Value(s)-based management, as we have defined it, represents a marriage between the shareholder-centric orientation of traditional VBM and the society-centric orientation of the CSR movement. We do not see the two concepts as mutually exclusive options or even substitutes for one another; rather, they are complementary. Value(s)-based management, we propose, provides a foundation for what has been called the *virtuous circle of corporate social responsibility*: doing well by doing good. Nonetheless, it does not follow that VBM is immoral and CSR is moral. Instead, values-based management is about using the two concepts to create wholeness or completeness.

Perhaps it is appropriate to provide a simple analogy of the various constructs that we discuss throughout this book. We argue that the ultimate goal of the corporation is to maximize long-term firm value. This can be represented by a pie; the bigger the pie, the greater the value. In order to measure how much value the pie represents and determine whether that value is increasing, we need a metric. In a later chapter we introduce economic value added (EVA) as such a metric. Economic value added can therefore be thought of as a scale that can weigh the pie. Of course, in order to bake the pie, a chef will find a cookbook to be very helpful since it provides a list of ingredients that need to be put into the pastry for it to be its best. Corporate social responsibility can be thought of as the cookbook that helps management operate in a socially responsible manner by considering the firm's various stakeholders and how they contribute to the firm's long-term sustainable value creation. Of course, the thing that holds this all together, the underlying foundation, is the chef, who chooses a particular recipe from the cookbook, ultimately gets a slice of the pie, and determines how big that slice will be. Value(s)-based management provides such a foundation and represents the bringing together of each of these elements—value creation, VBM, CSR, and EVA—into a cohesive management tool.

The distinction between VBM and CSR can be better understood by considering the philosophical underpinnings of each. The former sets as a goal the transformation of the cultural mindset within the firm to one of maximizing the firm's value. The latter goes further by attempting to create a cultural mindset that considers *how* to operate in order to create firm value. It also promotes the idea that firms should operate in a socially responsible manner such that the firm's impact on each of its stakeholders is considered, including the way in which each stakeholder's involvement can promote long-term sustainable value creation.

Appendix 1A

John Mackey and Milton Friedman on the Goal of the Firm

The debate between shareholder value and stakeholder theory becomes quite tangled when one considers the case of Whole Foods Market and its founder, CEO John Mackey. Perhaps "tangled" is not the correct word; a better word may be "irrelevant." Whole Foods demonstrates that, by making the pie bigger, everyone gets a bigger slice. The following conversation between John Mackey and Nobel laureate economist Milton Friedman appeared in the 2007 proxy statement for Whole Foods:

John Mackey:
I believe that the enlightened corporation should try to create value for *all* of its constituencies. From an investor's perspective, the purpose of the business is to maximize profits. But that's not the purpose for other stakeholders—for customers, employees, suppliers, and the community. Each of those groups will define the purpose of the business in terms of its own needs and desires, and each perspective is valid and legitimate.

My argument should not be mistaken for a hostility to profit. I believe I know something about creating shareholder value. When I co-founded Whole Foods Market 27 years ago, we began with $45,000 in capital; we only had $250,000 in sales our first year. During the last 12 months we had sales of more than $4.6 billion, net profits of more than $160 million, and a market capitalization over $8 billion.

But we have not achieved our tremendous increase in shareholder value by making shareholder value the primary purpose of our business.... The most successful businesses put the customer first, ahead of the investors. In the profit-centered business, customer happiness is merely a means to an end: maximizing profits. In the customer-centered business, customer happiness is an end in itself and will be pursued with greater interest, passion, and empathy than the profit-centered business is capable of.

Not that we're only concerned with customers. At Whole Foods, we measure our success by how much value we can create for all six of our most important stakeholders: customers, team members (employees), investors, vendors, communities, and the environment....

Many thinking people will readily accept my arguments that caring about customers and employees is good business. But they might draw the line at believing a company has any responsibility to its community and environment. To donate time and capital to philanthropy, they will argue, is to steal from the investors. After all, the corporation's assets legally belong to the investors, don't they? Management has a fiduciary responsibility to maximize shareholder value; therefore, any activities that don't maximize shareholder value are violations of this duty. If you feel altruism towards other people, you should exercise that altruism with your own money, not with the assets of a corporation that doesn't belong to you.

This position sounds reasonable. A company's assets do belong to the investors, and its management does have a duty to manage those assets responsibly. In my view, the argument is not *wrong* so much as it is too narrow.

First, there can be little doubt that a certain amount of corporate philanthropy is simply good business and works for the long-term benefit of the investors. For example: In addition to the many thousands of small donations each Whole Foods store makes each year, we also hold five 5% Days throughout the year. On those days, we donate 5 percent of a store's total sales to a nonprofit organization. While our stores select worthwhile organizations to support, they also tend to focus on groups that have large membership lists, which are contacted and encouraged to shop our store that day to support the organization. This usually brings hundreds of new or lapsed customers into our stores, many of whom then become regular shoppers. So a 5% Day not only allows us to support worthwhile causes but is an excellent marketing strategy that has benefited Whole Foods investors immensely.

That said, I believe such programs would be completely justifiable even if they produced no profits and no P.R. This is because I believe the entrepreneurs, not the current investors in a company's stock, have the right and responsibility to define the purpose of the company. It is the entrepreneurs who create a company, who bring all the factors of production together and coordinate it into viable business. It is the entrepreneurs who set the company strategy and who negotiate the terms of trade with all of the voluntarily cooperating stakeholders—including the investors.

Another objection to the Whole Foods philosophy is where to draw the line. If donating 5 percent of profits is good, wouldn't 10 percent be even better? Why not donate 100 percent of our profits to the betterment of society? But the fact that Whole Foods has responsibilities to our community doesn't mean that we don't have any responsibilities to our investors. It's a question of finding the appropriate balance and trying to create value for all of our stakeholders.

Milton Friedman:

The differences between John Mackey and me regarding the social responsibility of business are for the most part rhetorical. Strip off the camouflage, and it turns out we are in essential agreement. Moreover, his company, Whole Foods Market, behaves in accordance with the principles I spelled out in my 1970 *New York Times Magazine* article.

With respect to his company, it could hardly be otherwise. It has done well in a highly competitive industry. Had it devoted any significant fraction of its resources to exercising a social responsibility unrelated to the bottom line, it would be out of business by now or would have been taken over.

Here is how Mackey himself describes his firm's activities:

1) "The most successful businesses put the customer first, instead of the investors" (which clearly means that this is the way to put the investors first).

2) "There can be little doubt that a certain amount of corporate philanthropy is simply good business and works for the long-term benefit of the investors."

Compare this to what I wrote in 1970:

"Of course, in practice the doctrine of social responsibility is frequently a cloak for actions that are justified on other grounds rather than a reason for those actions.

"To illustrate, it may well be in the long-run interest of a corporation that is a major employer in a small community to devote resources to providing amenities to that community or to improving its government....

"In each of these...cases, there is a strong temptation to rationalize these actions as an exercise of 'social responsibility.' In the present climate of opinion, with its widespread aversion to 'capitalism,' 'profits,' the 'soulless corporation,' and so on, this is one way for a corporation to generate goodwill as a by-product of expenditures that are entirely justified in its own self-interest.

"It would be inconsistent of me to call on corporate executives to refrain from this hypocritical window-dressing because it harms the foundations of a free society. That would be to call on them to exercise a 'social responsibility'! If our institutions and the attitudes of the public make it in their self-interest to cloak their actions in this way, I cannot summon much indignation to denounce them."

Finally, I shall try to explain why my statement that "the social responsibility of business [is] to increase its profits" and Mackey's statement that "the enlightened corporation should try to create value for all of its constituencies" are equivalent.

Note first that I refer to *social* responsibility, not financial, or accounting, or legal. It is social precisely to allow for the constituencies to which Mackey refers. Maximizing profits is an end from the private point of view; it is a means from the social point of view. A system based on private property and free markets is a sophisticated means of enabling people to cooperate in their economic activities without compulsion; it enables separated knowledge to assure that each resource is used for its most valued use and is combined with other resources in the most efficient way.

Chapter 2

The Elements of Value-Based Management

> It is easy to forget why senior management's most important job must be to maximize its firm's current market value. If nothing else, a greater value rewards the shareholders who, after all, are the owners of the enterprise. But, and this really is much more important, society at large benefits too. A quest for value directs scarce resources to their most promising uses and most productive uses. The more effectively resources are deployed and managed, the more robust economic growth and the rate of improvement in our standard of living will be. Adam Smith's invisible hand is at work when the investor's private gain turns into a public virtue. Although there are exceptions to this rule, most of the time there is a happy harmony between creating stock market value and enhancing the quality of life.
>
> —G. Bennett Stewart III, *The Quest for Value*

Chapter 1 demonstrates the notion that a firm's management should make decisions that lead to increased long-term firm value, implying increased shareholder value, while also improving the economic well-being of the firm's employees, suppliers, and customers and of the communities in which the company operates. In fact, it is not uncommon to see corporate mission statements that endorse the invisible hand concept, which creates shareholder value while improving society as a whole. For example, the 2007 annual report of the Briggs and Stratton Corporation states: "We will create superior value by developing mutually beneficial relationships with our customers, suppliers, employees and communities.... In pursuing this mission, we will provide power for people worldwide to develop their economies and improve the quality of their lives and, in so doing, add value to our shareholders' investments." What this mission statement recognizes is that the only way to ensure the maximization of long-term shareholder value is to pay attention to the needs and interests of all of the firm's stakeholders. Likewise, values(s)-based management states that the motive for focusing on the interests of stakeholders other than shareholders is that this is good for the shareholders' interests.

While the primary goal of any company is to manage its operations to create long-term firm value, the incentives of a firm's management are not always aligned with those of the company's stockholders. As a consequence, many (perhaps even most) large corporations are not run on a day-to-day basis so as to maximize the firm's value. In fact, many may unintentionally destroy value year after year.

Wealth Creation Is Not Universal

Table 2.1 contains a list of the top five and bottom five wealth creators among the one thousand largest U.S. corporations as of year-end 2005. The ranking is based on market value added (MVA), a term coined by Stern Stewart and Company to measure how much wealth a firm has created at a particular moment in time. Market value added is equal to the difference in a firm's market value, both debt and equity, and the amount of capital that has been invested in the company. The names of the firms at the top of the list are probably very familiar to you. For example, investors have invested roughly $125 billion in the assets of top-ranked General Electric, whose market value at the end of 2005 was more than $407 billion. In other words, the market value added to the original investment is more than $282 billion at the end of 2005. At the other end of the spectrum we see JDS Uniphase, whose investors have entrusted more than $47 billion to the firm's management, is worth almost $41 billion less than their investment at the end of 2005. Table 2.1 also provides information on the rates of return earned on each firm's invested capital, as well as the firm's cost of capital. These last two pieces of information highlight a fundamental paradigm of value-based management: Firms that earn rates of return that are higher than their capital costs create shareholder wealth, whereas those that fail this simple test destroy it.

Table 2.1. America's Greatest Creators and Destroyers of Shareholder Wealth ($ in Millions)

Company Name	Invested Capital	Market Value	Market Value Added	Return on Invested Capital	Cost of Capital
Top 5 Wealth Creators in 2005					
General Electric	124,960	407,505	282,545	12.2%	6.9%
Microsoft	28,159	257,190	229,031	40.9%	11.7%
Proctor & Gamble	50,270	223,004	172,734	15.8%	7.2%
Exxon Mobil	229,608	397,167	167,560	18.9%	6.1%
Wal-Mart	109,393	250,098	140,705	10.8%	5.8%
Bottom 5 Wealth Creators in 2005					
Pfizer	209,293	189,867	(19,427)	5.8%	7.6%
Time Warner	132,985	109,790	(23,195)	3.8%	7.8%
AT&T	192,158	162,765	(29,393)	3.1%	8.8%
Lucent Technologies	61,987	25,422	(36,566)	–0.7%	9.6%
JDS Uniphase	47,506	6,896	(40,610)	–0.8%	11.8%

How do some firms create so much value while others destroy it? Value creation, very simply, results from the marriage of opportunity and execution. Opportunities must be recognized (and in some cases created), and this is the stuff of which business strategy is made. However, opportunity is not enough. Firms have to have employees that are ready, willing, and able to take advantage of business opportunities, and it is this side of the value creation equation that this book focuses upon. Specifically, how can we design a system of incentives that encourage employees to think and act like business owners?

The Proper Design of a V_sBM Program

Financial economists, beginning with Berle and Means (1932), have addressed the fundamental problems that arise when ownership and control of the modern corporation are separated. An agency problem occurs when a firm's owners, the stockholders, are different from its management. Fundamentally, managers control the firm and can make decisions that benefit themselves at the expense of the firm's stockholders. The various proponents of value-based management systems believe they have the answer to this problem. Each of the many VBM consultants would have you believe that using their proprietary VBM metric is the best way to measure the contributions of individuals and groups toward the creation of shareholder value. We find that, while each of these metrics may appear different on the surface, they all share certain key elements needed in the process of creating value. Regardless of the VBM consultant relied upon (or the VBM metric utilized), the process of value creation is the same.

Figure 2.1 captures the process of a V_sBM system designed to build and support a sustainable cycle of value creation. We emphasize the notion of sustainability; after all, value is created over time as a result of a continuing cycle of strategic and operating decisions. The fundamental premise upon which V_sBM systems are based is that, to sustain the wealth-creation process, managerial performance must be measured and rewarded using metrics that can be linked directly to the creation of a firm's value. Thus, the integration of value-based performance metrics and incentive compensation is at the very heart of V_sBM programs. Designing and implementing such a program is the challenge of V_sBM. In doing so, we have found that five primary elements are essential to the success of such a program:

1. *The V_sBM program must have full and complete support of the company's top executives.* Very simply, successful systems are top-down directives that, in many cases, completely transform the firm's

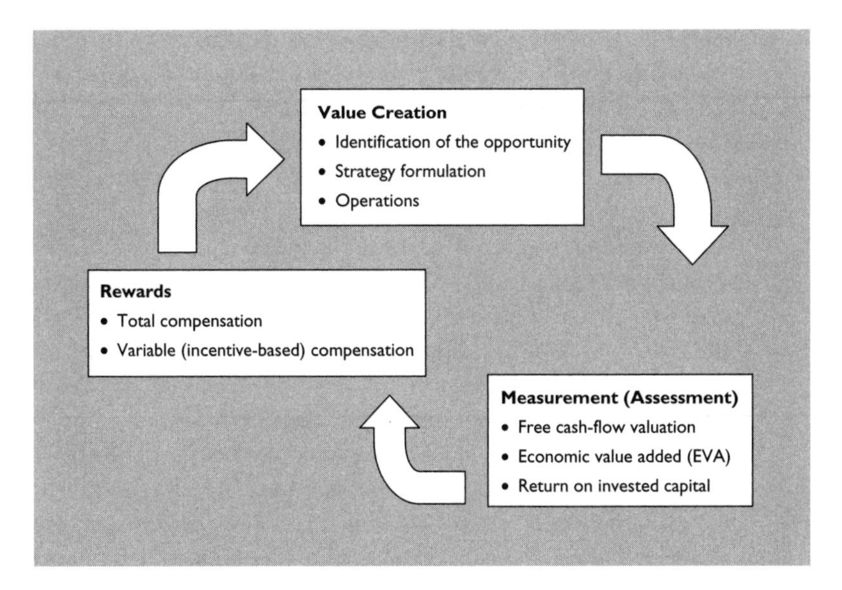

Figure 2.1. Constructing a sustainable cycle of value creation

operating culture. Although the impetus for the adoption of the system may have come from a planning group, financial officer, or someone else in the firm's corporate hierarchy, it is essential that the program gain the support of the CEO if it is to have a reasonable chance of success.

2. *For the V_sBM program to impact individual manager behavior, there must be a link between behavior and compensation.* This is a straightforward restatement of the old adage "What gets measured and rewarded gets done."
3. *Employees must understand the V_sBM system if it is to be effective in transforming behavior.* This frequently means that simplicity is often preferred to finely tuned measurement.[1] Simply put, systems work best when the firm's employees understand and accept the basic premise of the system and are able to implement it in their day-to-day discussions. Thus, education and training are absolutely essential to the future success of any V_sBM program. Value(s)-based management is about transforming behavior, and for any program to be successful, employees must understand what they are being asked to do, why it is important, and how their personal well-being will be impacted.
4. The measurement-and-reward system must be *capital-market focused.* That is, the key to successfully transforming an employee who thinks like an employee into one who thinks like an owner is to measure and reward the employee using methods that

parallel the rewards earned by the owners. This requires that the firm's internal measurement-and-reward system mirror the external capital-market system as closely as possible.

5. *Perhaps most important is that managers realize that the V_sBM metric is only a measurement of success; it is not in itself success.* Managers must be careful not to become so narrowly focused on the metric that they lose sight of what drives the metric. In other words, the metric tells you the score, not how to score. The organization must be so managed as to ultimately lead to improvements in the V_sBM metric, and to do so, managers must be careful not to take their eyes off the drivers of success. Firms are successful when they are able to meet their customers' needs by delivering goods and services at a profit. Value(s)-based management programs are implemented to facilitate this process by providing incentives that make employees act like owners. It is critical, however, to remember that it is delivery of value to the customer—and not the metric itself—that leads to shareholder value.

These five elements surface on numerous occasions in the chapters that follow.

Alternative Valuation Paradigms: Earnings versus Discounted Cash Flow

Two competing paradigms have been used to measure the creation of a firm's value: the earnings model and the discounted cash-flow model. Figure 2.2 captures the essential elements of both.

Although both of these models can be consistent in theory, they are not generally used in a consistent manner. If management uses the accounting model to think about value, it will focus on reported

	Earnings Model	Discounted Cash-Flow Model
Equity Value	(*Price Earnings Ratio*) x (*Earnings per Share*)	*Present Value of Future Cash Flows*
Value Drivers	Determinants of accounting earnings and the price/earnings ratio	Determinants of firm future cash flows and the opportunity cost of capital

Figure 2.2. Competing models of firm valuation

earnings in conjunction with the market's valuation of those earnings as reflected in the price/earnings ratio. For example, if the price/earnings ratio is 20, then a dollar increase in earnings per share will create $20 in additional equity value per share. Similarly, a $1 loss in earnings per share will lead to a drop of $20 in share value.

To see what is wrong with this assessment of value, consider the following scenario. A firm invests $1 in current-period earnings per share in research and development, which is expected to create valuable investment opportunities for the firm in the future. According to generally accepted accounting principles (GAAP), any R&D expenditures must be expensed in the current period, even though they are expected to create valuable investment opportunities in the future. In this circumstance investors may not penalize the firm's share price and may even drive it to a higher level following the announced investment in R&D despite the lower earnings per share.[2]

The discounted cash-flow model of valuation incorporates investors' expectations of cash flows into the indefinite future, as well as the opportunity cost of funds when determining the company's value. In this model, the R&D investment used in the previous example would lead to a reduction in cash flow during the periods in which the investment is being made but would correspondingly increase future cash flows when the anticipated rewards of the investment are being reaped.

New capital investment or capital budgeting analyses in virtually all firms are now based upon a discounted cash-flow model, which is consistent with the basic notion that a firm's stock value at any moment is equal to the discounted value of expected future cash flows accruing to the shareholders. However, these same firms often use earnings as the primary value driver to evaluate the performance of the capital that is already in the firm. The problem with using earnings in this way (as we have just seen) relates to the fact that maximizing earnings and earnings growth does not necessarily maximize share value since share value reflects the present value of all future cash flows, not just current earnings. To redress this problem, VBM models use performance metrics that are based upon discounted future cash flows.

Connecting Business Strategies with the Creation of Firm Value

How can we know whether a strategy will create value? In an uncertain world the answer to this question can be known only after the strategy has been implemented and run its course. However, a manager of operations must assess the success or failure of the strategy at finite intervals along the way so that those responsible can be rewarded for their success or punished for their shortcomings. The typical metrics used to measure a firm's periodic performance are based on historical accounting information that is readily available. Examples include firm earnings, earnings growth, or financial ratios such as the firm's profit margin or return on invested capital. These are almost always single-period, accounting-based measures of performance that suffer from two important limitations: First, since these performance measures are based solely on one historical period of operations, there is no reason to believe that they are good indicators of value to be created over the entire life of the venture. Second, accounting information systems do not incorporate an opportunity cost of the owner's capital. It is these limitations that V_sBM systems attempt to overcome.

Summary

The evidence clearly shows that a significant number of managements have not created firm value but instead destroyed it. The list of value destroyers includes a host of "respectable" firms. The cause of this damage has been associated largely with an agency problem created by the separation of the firm's management and its owners (stockholders). The shareholders frequently came to believe that many managers were working in their own best interest and not in the shareholders'—and they were none too happy about it. In the 1980s and 1990s this fact contributed significantly to shareholder activism and prompted a strong interest in value-based management metrics.

In the design of value-based metrics, there has been a long-standing debate about what matters. Essentially, two paradigms exist: the accounting (earnings) model and the discounted cash-flow model, and the crucial question is, what drives value—earnings or cash flow? While both the business press and managements have continued to focus on earnings, most VBM metrics are based on the discounted cash-flow model in an attempt to overcome the problems inherent in the accounting model of firm valuation.

In the early days of the VBM movement, the primary question was, who has the best metric for measuring value creation? Vendors who were selling proprietary software to measure a firm's value traveled around the country telling prospective clients why their software was better than the competition's. However, it soon became apparent that simply having a metric was not sufficient; a system needed to be designed to build and support a sustainable cycle of value creation, including a process that connected the firm's business strategies to the creation of firm value. At a minimum such a system required (1) the full support of the top executives; (2) a link between behavior and compensation; (3) the employees' understanding of the system; (4) capital market focus; and (5) the realization that a VBM metric is only a measure of success—not success itself.

In this chapter we have introduced the concept of value-based metrics and the need to overcome the problems inherent in the accounting model of valuation. We continue this discussion in the chapters that follow.

Chapter 3

The Need to Measure What You Want to Manage

> Data, metric, measures, assessments, evaluations, scorecards, progress reports...Many of us have been faced with a whole host of measurement opportunities. Seems like some of the performance measurements are moving targets that we seldom hit. Some measurement processes come and go like fog. I recall hearing this maxim years ago: "What gets measured gets done." It has been attributed to Peter Drucker, Tom Peters, Edwards Deming, Lord Kelvin, and others. Why is it that so often we get hung up on metrics and measuring things to the point that we sometimes lose track of measuring what really matters?
>
> —Robert M. Williamson, "What Gets Measured Gets Done: Are You Measuring What Really Matters?"

Perhaps no adage in management rings truer than the doctrine that you cannot manage something that you cannot measure. It is also often said that what you don't know won't hurt you, but nothing could be further from the truth when you are running a business. The concept is really quite simple—unless you measure something, you do not know whether it is getting better or worse. You cannot manage for improvement if you do not measure to see what is getting better and what is not. However, while the concept may be quite simple, the devil, as with many things, is in the details. What, within the area of value-based management, should we measure? Just what metric captures what we wish to manage? These are not simple questions, as the countless examples of inappropriate choices demonstrate.

The Need for a Single Metric

Before we venture too far down the path of answering these questions, let us first pause and discuss the reasons we will be focusing our measurement on shareholder value rather than on the multiple metrics required under stakeholder theory.

The underlying problem with stakeholder theory comes down to basic mathematics. It is not possible to maximize more than one item at a time (e.g., profits, quality, market share, efficiency) unless they are all simple transformations of one other. Consequently, it is essential, under this multiple measures formula, to specify tradeoffs between the multiple dimensions so that the manager has some guidance in making an informed choice.

Jensen (2001) argues that stakeholder theory should not be viewed as a legitimate contender for value maximization since it fails to provide managers a means to balance the conflicting demands of a corporation's various stakeholders. Customers want high-quality goods and great services at low prices; employees want high compensation and a stress-free work environment; communities want significant social contributions; the government wants high tax receipts, and so on. The manager is left without decision criteria by which to make the necessary tradeoffs. In contrast, value maximization clearly states that the manager should provide as many resources as each stakeholder demands as long as the benefits received (i.e., the long-term value created) exceed the additional costs. In other words, value maximization offers a way out of this conundrum by supplying the manager with a decision rule. In addition, this directive, as previously demonstrated, is the one that adds the most value to society.

One effort to overcome the inadequacy of stakeholder theory has been the *balanced scorecard*, which is intended as a way to use a firm's strategy to formulate its objectives. The basic thought behind the scorecard is that a single measure (typically financial) cannot provide a complete picture of the organization. Financial measures are by their very nature backward looking, indicating what has already happened. In contrast, a firm requires forward-looking metrics, so-called leading indicators, in order to properly determine managerial actions. For example, a company may choose metrics that supply indicators not only of financial metrics but also concerns such as process efficiency, customer satisfaction, and employee morale. The overall objective of the scorecard is to include a set of metrics that allows managers to monitor the entire business.

The balanced scorecard has the potential to provide a management framework that can help managers track the many factors that influence performance. Unfortunately, like stakeholder theory, the balanced scorecard lacks a single metric to enable management to make tradeoffs. Managers need one overriding metric to tell them the ultimate score, and this metric must summarize the interaction between the indicators so that a manager has guidance when taking action. Fortunately, a proper VBM metric included within a balanced scorecard can provide such a metric. Robert Kaplan, one of the creators of the balanced scorecard, stated: "Creating EVA [a popular VBM metric discussed in chapter 5] is the ideal outcome of a successful strategy, and that's what we are trying to do."[1]

Greg Milano, a former partner at Stern Stewart, explains the balanced scorecard and EVA in terms of soccer:

> In a sense, business is like football. A manager of a football club needs to push many factors to succeed, such as getting shots on goal, winning corner kicks, providing a solid defense and having a goalie who makes saves. At the end of the match, however, all that matters is that you win, not how many corner kicks you get. In business too, a manager must manage many inputs, and both ABC [activity-based costing] and the balanced scorecard help managers to make decisions. But EVA is needed to determine whether we win or we lose. By using EVA for decisions, performance measures and rewards, managers are motivated to use the information at hand to act as owners and create value.[2]

To summarize, we maintain that the basic tenets of corporate social responsibility and VBM can be combined by using a balanced scorecard device as a very useful tool for creating shareholder value. In this context, CSR and value maximization are viewed as complements, not substitutes. Value maximization provides a simple decision rule for making tradeoffs but does not provide a recipe for how to add value. Corporate social responsibility helps in that it identifies all of the relevant stakeholders and emphasizes their importance much as the cookbook does in the analogy we used in chapter 1. It is absolutely essential for long-term survival that a firm consider the needs of all of its stakeholders.

As Peter Drucker has stated, the customer provides the foundation of the business and keeps it in existence. Moreover, without employees the firm cannot meet the customers' needs (Drucker 1954/1993, 39–40). Similar arguments can be made with regard to each stakeholder group. Consequently, stakeholder theory supplies a recipe for adding value, whereas value maximization theory provides a decision rule for determining the proportions of the various ingredients needed to maximize shareholder value.

Accepting the fundamental notion that a single metric is needed to measure success, we turn our attention to finding that metric. Many metrics have been advocated and used; however, most of them have serious shortcomings. Surprisingly, many of the traditional ones that large numbers of firms continue to use are fraught with problems. Before we introduce a measure that we feel can best guide value-based management, we discuss several traditional metrics that we believe should be abandoned.

Without a doubt, one of the most important components of a successful value-based management approach is the selection of a proper metric to keep score. Unfortunately, this task has proven elusive to many companies. The options that have been traditionally used can be grouped into market-based metrics or accounting-based measurements.

Most individuals believe that the best indicator of successful wealth creation is either total shareholder return or total market value, both of which are market-based metrics. Further, these same people will likely contend that the best way to attain total shareholder return or total market value is either through growth in a profit measure such as net income or earnings per share or through another accounting-based measure such as return on invested capital. Truth be told, all of these traditional market metrics are seriously flawed for the purpose of guiding a value-based management program. Total shareholder return can be a misleading measure of performance;[3] market value by itself is nearly useless; and profit measures such as net income and return on invested capital can actually lead to wrong decisions.

Total Shareholder Return

Total shareholder return would seem to be an excellent metric of value creation. After all, isn't it stock returns that shareholders seek when they invest their capital? Unfortunately, total shareholder return is deficient as a useful metric for the following reasons.

First of all, shareholder return involves what is known as the *controllability principle,* which holds that managers should be held accountable only for things that they can control. Shareholder return is influenced by management's actions; however, many items outside of its control also impact stock returns. Examples include economy-wide interest rates, general economic conditions, governmental actions, and the weather. Because of all these uncertainties, it is hard to determine whether a high return is due to management's savvy actions or simply luck.

Second, even if management's actions largely accounted for shareholder return, the metric could still be a misleading indicator of performance. While higher returns, which comprise price appreciation and dividends, are better than lower returns, this metric does not consider risk. Investors seek not simply higher returns but also higher risk-adjusted returns. Just because one stock has a higher return than another does not necessarily mean it outperformed the other stock—unless we adjust for risk. For example, a firm such as Amgen (in the high-risk biotech industry) would need to provide a much higher return than a firm such as Sempra (in the low-risk, regulated-utility industry) to compensate for the additional precariousness of holding the stock.

Further, we would need to consider not only the risk associated with industry-type effects but also that related to the amount of leverage a firm utilizes. All else being equal, a firm with a high financial leverage is riskier and requires a higher total shareholder return than a similar firm with low financial leverage.

Finally, total shareholder return calculations assume that all dividends are reinvested in the firm. Obviously this is not realistic overall since it would require some shareholders to sell their stock so that the other dividend-receiving shareholders could reinvest.

Total Market Value

Although total shareholder return is flawed as a metric for a value-based management program, it is better than total market value or market capitalization. Market value by itself tells us basically nothing about wealth creation. It merely explains how large a company is, not how it became large. Missing is how much capital went into making the company the size that it is. Wealth creation is the difference between the capital invested in the firm and the value of the resulting firm. A comparison of Time Warner and Coca Cola serves as a good example of the problem of considering only market value. Time Warner, the media giant, had a 2005 market value of almost $110 billion, while the giant beverage company Coca Cola was slightly smaller, with a 2005 market value of a little more than $102 billion. Based simply on market value,

Time Warner has outperformed Coca Cola. However, the amount of capital that has been invested in each company is not revealed. In fact, Time Warner has seen nearly $133 billion of capital invested and therefore has lost approximately $23 billion in value, whereas Coca Cola has had only $18 billion of capital invested and therefore has added approximately $84 billion in value. Thus, although Time Warner is larger, Coca Cola has been the superior performer.

Accounting-Based Metrics

We have seen that the two primary market-based metrics just described are not suitable metrics for managing shareholder value, but how about the traditional accounting-based measures, such as earnings or accounting return on invested capital? At least they pass the controllability requirement. Just by reading the financial press, one easily gets the impression that a firm's earnings per share are the principal driver of its stock price. For example, on October 22, 2007, Apple released an after-hours earnings announcement for its fiscal 2007 fourth quarter. The company reported that profits had increased 67% during the quarter and had earned $1.01 per share. These profits easily beat analysts' expectations of $0.86 per share. Apple's shares were one of the most actively traded the next day, gaining $11.80 per share, or 6.7%, on more than 64 million shares. It would appear that share value is directly tied to earnings and earnings growth.

Does a firm that is managed for earnings and earnings growth maximize the firm's value? Proponents of VBM and financial economists argue that a firm's earnings provide an insufficient indicator of value creation. Specifically, rewarding a firm's management for earnings and earnings growth can even lead to decisions that destroy rather than create value.

Most likely you have heard a story or two about accountants. Most of these tales help perpetuate certain beliefs, such as accountants' unwillingness to venture very far outside the box. Not many professions are known to be much more resistant to change, and this failure to change with the times has led some to claim that the information we get from our accounting system is no longer capable of providing the needed information to drive company value creation in the new economy. The following story illustrates this sentiment well.

A man takes a hot-air balloon ride at a local country fair. A fierce wind suddenly kicks up, causing the balloon to leave the fairgrounds and carry its occupant out into the countryside. Landing in a farmer's field next to a road, the man has no clue how far he has flown or where he is. Seeing a man walking down the road, he cries out: "Excuse me, sir, can you tell me where I am?" Eyeing the man in the balloon, the passerby says, "You are in a downed balloon in a farmer's field." "You must be an accountant, sir," replies the balloon's unhappy traveler. "How could you possibly know that?" asks the passerby. "Because what you have told me is absolutely correct but of absolutely no use to me now," answers the balloonist. The accountant then replies, "You must be a manager." The balloonist says, "How would you know that?" The accountant replies, "Because you don't know where you are, you don't know how you got here, you don't know where you're going, and you are exactly where you were ten minutes ago, but somehow it's now my fault!"

The story actually makes two points. First, in its quest for numbers that can be reliably verified, the current accounting system has sacrificed relevance. Some accounting information, as we will learn, may have a great deal of relevance to the decision maker but does not fit well into the accountant's system of recording. Consequently, the information is ignored. The second point is that we should not blame the accountants for this situation. Managers need to realize the limitations of their traditional accounting systems and develop better metrics to help them to determine where they are and how to get to where they want to go.

Recounting what we have learned so far, one accepted belief in management is that you get what you measure and reward. Furthermore, if you fail to measure something, you will likely achieve it only through luck. Rewards that are tied to a measurement can be a very strong motivator and are not something that should be treated lightly. It is critical that the correct thing be measured for the desired outcome; otherwise, you may end up motivating entirely unexpected behavior.

Think Economic Profits, Not Accounting Profits

Our present financial accounting system comprises three major statements: the balance sheet, the income statement, and the statement of cash flows. The output from these records allows us to compute many potential metrics. At a particular point in time, the balance sheet depicts an entity's financial position, which is determined as assets less liabilities. The income statement presents the entity's performance for a period of time measured in terms of accrual revenues and expenses, yielding a "bottom-line" net income. Finally, the statement of cash flows identifies sources and uses of cash and represents a cash-basis measure of performance for a period of time. While many feel that it is cash flow that ultimately matters, generally accepted accounting principles (GAAP) are far more concerned with the accrual definition of performance and financial position as reported on the income statement and the balance sheet. Two widely used measures of performance are net income and return on invested capital (ROIC).

Assume for the moment that you have $5,000 to invest but that you are able to find only two new investment opportunities. One opportunity requires $1,000 with the constraint that no additional amount is allowed to be invested in this venture. The second opportunity requires the entire $5,000 to be invested. Table 3.1 reports two performance metrics for each investment. For this example, assume that these results are known in advance with perfect certainty, so there is no risk involved.

Which of the investment opportunities presented in table 3.1 represents the better investment? This is really a trick question that highlights the problem with traditional accounting metrics. Some may look at the higher net income generated by investment B and claim this to be the better investment. They may argue that net income is the bottom-line definition of performance. Others, however, may counter that net income fails to consider the balance sheet and therefore ignores the investment needed to earn the reported net income. A much larger investment should earn more net income. Investment A, with its higher reported ROIC, must therefore be the better investment.

Unfortunately, neither net income nor ROIC considers what you gave up earning by choosing one of these two investment opportunities. The term economists use for this concept is *opportunity cost,* the cost of your next best opportunity. For the moment, let us assume that you had been earning 11% on your money before making one of these two investments. Therefore, you give up (or have an opportunity cost of) $110 for every $1,000 you invest. In Case 1, table 3.2 introduces a new metric, *economic profit,* which takes into account this opportunity cost.[4]

Table 3.1. Invested Capital, Net Income, and ROIC for Two Investment Opportunities

	Investment A	Investment B
Invested capital	$1,000	$5,000
Net income	$150	$500
ROIC	15%	10%

Table 3.2. Case 1: Opportunity Cost of Capital = 11%

	Investment A	Investment B
Invested capital	$1,000	$5,000
ROIC	15%	10%
Net income	$150	$500
Capital charge (i.e., the opportunity cost of capital)	$110	$550
Economic profit	$40	$(50)

Table 3.3. A Summary of the Potential Investment Outcomes

	Status Quo	Investment A	Investment B
$5,000 @ 11%	$550		
$1,000 @ 15%		$150	
$4,000 @ 11%		$440	
$5,000 @ 10%			$500
Total	$550	$590	$500

Economic profit is shown as the profit that remains after considering the opportunity cost of invested capital. As table 3.2 illustrates, after considering the opportunity cost of 11%, only investment A has a positive economic profit. This is because only investment B earns a return on the invested capital that is greater than the 11% that could otherwise have been earned. This is more clearly reflected in table 3.3.

Table 3.3 shows that you are better off investing in A since the total return from your $5,000 is the highest of the three options, one of which is the status quo (i.e., leaving all of your money earning 11%). Investment B, in contrast, destroys $50 of wealth relative to the status quo option.

Net income thus does not give us the information we need to make proper investment choices, something managers must do continually when determining how to run their companies. But what about ROIC? Why not just go with the investment opportunity that yields the highest ROIC? The problem is that ROIC, like net income, also fails to consider opportunity costs. What if the opportunity cost is only 7% rather than 11%? Case 2 in table 3.4 illustrates this situation.

Now things reverse, and investment B looks better than investment A in terms of economic profit. How can that be? Table 3.5 provides the computations to better illustrate this.

Table 3.4. Case 2: Opportunity Cost of Capital = 7%

	Investment A	Investment B
Invested capital	$1,000	$5,000
ROIC	15%	10%
Net income	$150	$500
Capital charge (i.e., the opportunity cost of capital)	$70	$350
Economic profit	$80	$150

Table 3.5. A Summary Performance of Investment Outcomes

	Status Quo	Investment A	Investment B
$5,000 @ 7%	$350		
$1,000 @ 15%		$150	
$4,000 @ 7%		$280	
$5,000 @ 10%			$500
Total	$350	$430	$500

Now both investments A and B are able to earn positive economic profit since both earn returns on invested capital in excess of the opportunity cost of the invested capital. Investment B is superior, however, because it is able to earn these excess returns on a larger amount of invested capital. Think in terms of an extreme example. How excited would you be to earn 1,000% on your investment if you could invest only $1? While 1,000% may look fantastic in terms of percentage returns, this investment adds only $10 to your wealth.

So neither net income nor ROIC gives us the information to make proper investment decisions in the way that economic profit can. In fact, net income and ROIC, if relied upon, can lead to some very bad wealth-destroying behavior. For example, investing in B in case 1 actually destroys wealth even though a positive net income is reported. This is a serious flaw with traditional GAAP accounting that fails to consider the opportunity cost of shareholder investments. While a charge against income for debt financing is present in the form of interest expense, no such charge exists for equity financing. In a sense, the accounting system treats equity capital as free. Shareholders that invest in a company, however, expect a return on their investment. The following example further illustrates how a reliance on net income, as well as the equally flawed metric of earnings per share, can lead to wealth-destroying behavior that is actually rewarded.

Consider a $1 million project that is predicted to produce operating returns of $75,000 per year into perpetuity. If we assume the firm has a 10% cost of capital, the perpetuity has a present value of $750,000. Therefore, after subtracting the initial investment of $1 million, the overall project has a negative net present value (NPV), $250,000. We could do a similar calculation using economic profit. The project will have a negative $25,000 annual economic profit calculated as $75,000 less a capital charge of $100,000 ($1 million times the firm's 10% cost of capital), or a negative NPV of $250,000. Using economic profit, it is apparent that this project will destroy value.[5]

Now look at how this project will look based on accounting earnings. Annual operating earnings (i.e., earnings before interest and taxes) are increased by $75,000 because of the project. Even after considering financing, the results will look impressive in most cases because most projects are financed with a combination of debt and equity (for example, with a mix of about two-thirds equity and one-third debt). The accounting system includes a charge for the cost of debt in the form of interest expense; however, no such charge exists for equity financing. If we use a 6% interest charge in this example, the firm will earn a before-tax income of $55,000 ($75,000 less interest expense of $1 million times one-third times 6%). In fact, as long as the project can earn a rate large enough to cover the interest charge (overall 2% in this example[6]), accounting net income will improve even though the project destroys value by earning well under the 10% overall cost of capital.

By now it should be clear that a reliance on net income alone (as defined by our current accounting definitions) is problematic, but will ROIC also lead to suboptimal behavior? The following example demonstrates that ROIC is also problematic.

For the sake of illustration, assume that Savanna Company is made up of two divisions. Both divisional managers are measured and rewarded based on the ROIC of each one's division, and greater levels of ROIC lead to increased manager compensation. The Turkey division earns an ROIC of 10%, while the Leader division earns an ROIC of 20%. Finally, assume the Savanna has an opportunity cost of capital of 15%, such that any project that earns greater than 15% increases Savanna's shareholder value, whereas projects that earn less than this amount destroy value.

What do you suppose will happen if both divisions become aware of a project that, according to the best information available, promises a

13% return? Obviously this project is not in Savanna's best interest and should be rejected. With bonuses based on the level of ROIC, Leader's manager will reject this project since it will lower Leader's overall ROIC and therefore lower the manager's bonus. The same cannot be said of Turkey's manager, however. Even at the expected below-cost-of-capital return of 13%, the Turkey division will show an increase of division ROIC with this new project. Therefore, it makes sense for Turkey's manager to paint a rosy picture when presenting this project for funding, perhaps by using overly optimistic assumptions to drive the portrayed ROIC above the 15% threshold.

Now consider another project that, according to the best information available, promises a 17% return. Savanna would likely want to fund this project, and Turkey division will most likely wish to have it. But what about the Leader division? A 17% return, while above the company's cost of capital, will lower the Leader division's ROIC and therefore not be in its best interest. You may be thinking that Savanna could simply require Leader to take this project, but that could occur only if Savanna knows about it. In many cases it is the divisions that become aware of potential projects and submit them for funding. Leader may simply never make Savanna aware of this project.

So here we have a metric (ROIC) that can provide incentives for a situation in which most of the funds flow to the Turkey divisions while the Leader divisions are starved for capital. This is just the opposite of what is good for Savanna.

Summary

This chapter first explains the importance of using a single metric to manage the process of creating firm value. While multiple-measure techniques such as the balanced scorecard can supply valuable information, they fail to provide managers with a means to make necessary tradeoffs. Only one item can be maximized at a time.

The question then becomes one of what metric best measures value creation. The chapter discusses several shortcomings of traditional metrics, both market based (e.g., shareholder return and market value) and accounting based (e.g., net income and return on invested capital).

Problems with market-based metrics include a lack of both controllability and risk considerations. Use of the accounting-based metric net income can lead to investing in (and retaining) projects that fail to cover even their opportunity cost of capital, especially if they are equity financed. This is because the accounting definition of net income fails to

Appendix 3A

Accounting versus Economic ROIC

Seasoned financial analysts will quickly point out that many of the shortcomings of accounting earnings that we have discussed in this chapter can be overcome by combining income statement and balance sheet information into financial ratios. Specifically, the *accounting* return on invested capital ($ROIC_{(ACC)}$) ratio has been widely used as a measure of financial performance. Although $ROIC_{(ACC)}$ represents an improvement over earnings alone, it, too, is a flawed indicator of shareholder value creation. In fact, Bierman (1988) identified the problems of using the conventional accounting return on net assets (a measure similar to ROIC to evaluate performance).

Let us begin our discussion by defining two measures of $ROIC_{(ACC)}$. The first uses net income as its numerator, while the second uses net operating profit after taxes (NOPAT):

$$ROIC_{(ACC1)} = \frac{\textit{net income}}{\textit{invested capital}} \tag{3A.1}$$

$$\begin{aligned} ROIC_{(ACC2)} &= \frac{\textit{net income} + \textit{interest } x\,(1 - \textit{tax rate})}{\textit{invested capital}} \\ &= \frac{\textit{net operating income } x\,(1 - \textit{tax rate})}{\textit{invested capital}} \\ &= \frac{\textit{net operating profit after tax}}{\textit{invested capital}} \end{aligned} \tag{3A.2}$$

The first measure, or $ROIC_{(ACC1)}$, in equation 3A.1 is often criticized for inconsistently comparing after-interest income to the total asset base, while the second seeks to rectify this shortcoming by including in the income figure the after-tax payment to the firm's creditors. The second ROIC measure (equation 3A.2) is internally consistent in comparing total assets to total earnings, and for this reason we use this version of ROIC in the discussion that follows. However, even this variant of ROIC suffers from problems that make its use as a tool for evaluating strategies and performance at either the business unit or the corporate level questionable.

The first difficulty that arises with the use of $ROIC_{(ACC2)}$ as a performance measure is that it reflects accounting income rather than cash flow. Since the value of a strategy or business unit depends on the amount, timing, and riskiness of future cash flows, the use of ROIC can provide misleading signals for decision making in business.

To illustrate the shortcomings of the accounting-based $ROIC_{(ACC2)}$ metric, let us define a discounted cash flow (DCF) version of the ROIC metric, $ROIC_{(DCF)}$, that properly captures the economic return to a project over a specified interval of time:

$$ROIC_{(DCF)} = \frac{cashflow + (PVyearend - PVbeginningofyear)}{PVbeginningofyear}$$

or

$$ROIC_{(DCF)} = \frac{cashflow + changeinpresentvalue}{PVbeginningofyear}, \quad (3A.1)$$

where the change in present value is a project's *economic* depreciation, as opposed to *accounting* depreciation.

The DCF return (sometimes referred to as the total shareholder return [TSR]) is the economic-rate-of-return equivalent of ROIC. For example, consider the following investment proposal: The Alpha Resale Company is considering an investment of $4,000 in a project that is expected to produce cash flows of $400, $800, $800, $1,600, and $2,500 over the next five years. Alpha estimates that its cost of capital is 12% such that the project's net present value (NPV) is zero:

$$NPV = \frac{\$400}{(1.12)^1} + \frac{\$800}{(1.12)^2} + \frac{\$800}{(1.12)^3} + \frac{\$1{,}600}{(1.12)^4} + \frac{\$2{,}500}{(1.12)^5} + (\$4{,}000) = 0$$

Table 3A.1 contains calculations of the traditional accounting-based definition of $ROIC_{(ACC2)}$ and the discounted cash-flow variant, $ROIC_{(DCF)}$, for each year of the project's five-year life. In panel A we calculate the $ROIC_{(DCF)}$ for each of the five years of the project's life. Since the project earns a zero NPV, it also earns an annual $ROIC_{(DCF)}$ equal to the firm's cost of capital. Panel B contains the $ROIC_{(ACC2)}$ calculations, which rely entirely on accounting income and book value numbers. A quick comparison of the $ROIC_{(DCF)}$ and the $ROIC_{(ACC2)}$ calculations reveals that $ROIC_{(ACC2)}$ is not a very useful indicator of the $ROIC_{(DCF)}$. For instance, $ROIC_{(ACC2)}$ ranges from –11.11% to more than 425% and bears no apparent relationship to the DCF return earned by the project nor does it indicate that the project is a zero NPV project.

Thus, we must conclude that the traditionally measured accounting $ROIC_{(ACC)}$ cannot generally be used as a replacement for $ROIC_{(DCF)}$.

Table 3A.1. $ROIC_{(DCF)}$, ROIC, and Project Evaluation

	Year 1	Year 2	Year 3	Year 4	Year 5
Panel A. Calculation of $ROIC_{(DCF)}$					
1. Cash flows	$400	$800	$800	$1,600	$2,500
2. Present value of the remaining cash flows	$4,000	$4,080	$3,770	$3,422	$2,233
3. Present value at end of year	$4,080	$3,770	$3,422	$2,233	$0
4. Change in present value during the year (3) – (2)	$80	($310)	($348)	($1,189)	($2,233)
5. Economic income (cash flow + change in value) or (1) + (4)	$480	$490	$452	$411	$268
6. $ROIC_{(DCF)}$ (5) ÷ (2)	12.00%	12.00%	12.00%	12.00%	12.00%
Panel B. Calculation of ROIC					
Cash flows	$400	$800	$800	$1,600	$2,500
Less depreciation	$800	$800	$800	$800	$800
Operating income	($400)	$0	$0	$800	$1,700
Book value, beginning of year	$4,000	$3,200	$2,400	$1,600	$800
Less depreciation	$800	$800	$800	$800	$800
Book value, end of year	$3,200	$2,400	$1,600	$800	$0
Average book value	$3,600	$2,800	$2,000	$1,200	$400
Return on assets (ROIC)	–11.11%	0.00%	0.00%	66.67%	425.12%

Appendix 3B

More Problems with Accounting-Based Metrics

The preceding examples are simple illustrations of the failings of traditional accounting-based metrics in a VBM program. What follows is a more detailed discussion of these shortcomings. Proponents of VBM maintain that accounting numbers that are prepared using generally accepted accounting principles are not designed to reflect value creation. Furthermore, earnings provide very noisy and sometimes misleading signals to the financial manager who seeks to maximize shareholder value. Before we examine the limitations of accounting earnings, let us first review the fundamental tenets of the basic discounted cash-flow valuation model. The value of any earning asset using the DCF model is a function of the amount, timing, and risk of expected future cash flows. Consequently, if the manager is to run the business in such a way as to maximize shareholder value, the performance-measurement system must capture all three of these fundamental determinants of value. In the discussion that follows we find that GAAP earnings fall short in five important respects when used to manage shareholder value.

Accounting Numbers Do Not Equal Cash Flows

This criticism is actually a bit misleading, for we can certainly get to cash by making appropriate adjustments to reported accounting numbers. In fact, the VBM metrics we introduce in the following chapters begin with reported accounting numbers and then make appropriate adjustments. The point is, however, that reported earnings are not equal to cash, and cash is the thing we are concerned about when trying to manage the creation of shareholder value.

The financial economist focuses on value creation where value is a function of the amount, timing, and risk of the cash flows that the firm produces. In contrast, reported income is calculated in a firm's income statement following the principles of accrual accounting. This means that revenues are recognized when the earning process is essentially complete, and expenses are then matched to the revenues they helped generate. Thus, a firm's earnings can deviate significantly from cash flow.

Accounting Numbers Do Not Reflect Risk

Reported accounting earnings do not reflect the riskiness of those earnings. That is, a firm's accounting system reports "what happened," not "what might have happened." As a consequence, nothing in reported earnings indicates anything about the riskiness of the firm's operations. Since risk is a principal determinant of the value of the firm's equity, this omission is critical. This same criticism can be levied at cash flow as well since it similarly does not reflect risk. Our point is that we cannot just use earnings or cash flow as the basis for managing for shareholder value.

Both the level of a firm's reported earnings and its variability from period to period are influenced by a combination of factors that are outside the firm's control (e.g., overall business conditions in the economy and in the firm's industry), as well as policy choices the firm's management has made (e.g., the firm's choice of operating and financial policies). To understand the impact of a firm's policy choices on earnings variability, consider the effect of financial policy on the volatility of a firm's reported earnings for the two firms included in table 3B.1. The unlevered firm uses no financial leverage, as its name implies, whereas the levered firm has borrowed half of its funds and pays 12% interest on its debt. Figure 3B.1 depicts the greater sensitivity of the levered firm's earnings per share (EPS) to changes in the level of operating income or earnings before interest and taxes (EBIT). Note that if EBIT were to increase from \$100 to \$200 (a 100% increase), the EPS for the unlevered firm would

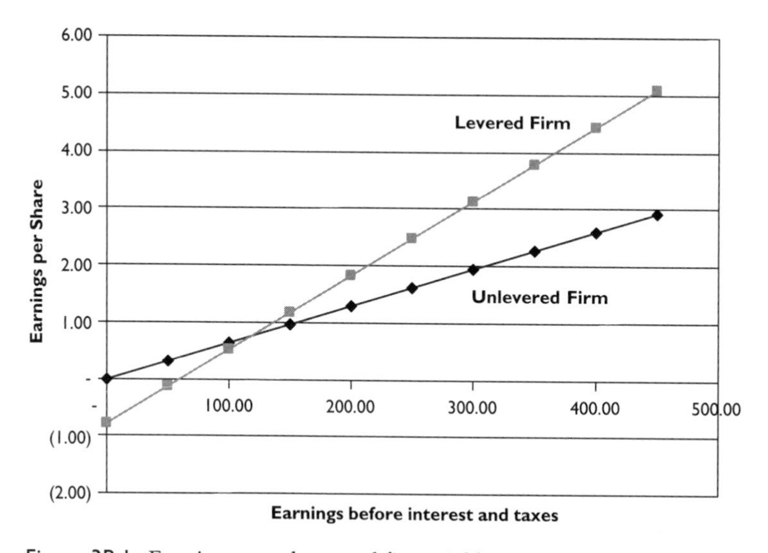

Figure 3B.1. Earnings per share and financial leverage

Table 3B.1. Financial Leverage and Earnings Volatility

	Unlevered Firm	Levered Firm
Debt	—	$500
Equity	$1,000	500
Interest rate	12%	12%
Tax rate	35%	35%
Number of shares outstanding	100	50

Unlevered Firm				Levered Firm			
EBIT	Interest	Net Income	EPS	EBIT	Interest	Net Income	EPS
—	—	—	—	—	60.00	(39.00)	(0.78)
50.00	—	32.50	0.33	50.00	60.00	(6.50)	(0.13)
100.00	—	65.00	0.65	100.00	60.00	26.00	0.52
150.00	—	97.50	0.98	150.00	60.00	58.50	1.17
200.00	—	130.00	1.30	200.00	60.00	91.00	1.82
250.00	—	162.50	1.63	250.00	60.00	123.50	2.47
300.00	—	195.00	1.95	300.00	60.00	156.00	3.12
350.00	—	227.50	2.28	350.00	60.00	188.50	3.77
400.00	—	260.00	2.60	400.00	60.00	221.00	4.42
450.00	—	292.50	2.93	450.00	60.00	253.50	5.07

increase by exactly 100%, that is, from $0.65 to $1.30, whereas the levered firm's EPS would increase by 350%, or from $0.52 to $1.82. Financial leverage has the effect of increasing the sensitivity of a firm's EPS to changes in EBIT, thus leading to a more volatile stream of earnings. This added volatility in EPS is not apparent in the reported earnings.

Accounting Numbers Do Not Contain an Opportunity Cost of Equity

As earlier chapters point out, the relationship between changes in economic value and earnings is further obscured by the fact that traditional earnings calculations take into account the opportunity cost associated with the owner's investment in the firm. Interest expense is considered when calculating accounting earnings, but no required return on equity capital is considered in the calculation of accounting earnings. The absence of a cost for owner-supplied capital means that reported earnings overstate the firm's value creation for the period in question. For example, in 2004 General Motors Corporation reported a net income of more than $2.8 billion. To many investors this positive earnings performance might have been interpreted as an indication that the firm was in good financial health. However, General Motors' total invested capital for the year was approximately $90 billion. If the company's investors require a 10% return on their capital, then the firm needs roughly $9 billion just to provide the required return. Obviously General Motors fell short of this mark by a substantial amount.

Accounting Practices Vary from Firm to Firm

Any student of financial statement analysis is aware of the impact (sometimes material) that a change in accounting policy can make on a firm's reported earnings. Typical examples include the various methods that a firm can use to account for its inventories (e.g., LIFO versus FIFO inventory) and the rules of governing accounting bodies such as the Financial Accounting Standards Board (FASB), which regulate the way in which a firm can account for R&D expenditures or foreign-currency translation gains and losses. Except for the impact on cash flow (e.g., via the firm's current and future tax liabilities), these accounting practices are unimportant insofar as the performance of the firm is concerned. Nonetheless, accounting practices can and do have a material influence on a company's reported earnings.

Accounting Numbers Do Not Consider the Time Value of Money

Reported earnings are not adjusted for the effects of the time value of money. Again, this same criticism can also be levied at cash flow. Economic value, on the other hand, incorporates consideration of the timing, as well as the amount and riskiness, of future cash flows. Specifically, a firm's economic or intrinsic value is equal to the present value of its expected future cash flows discounted at a rate that properly reflects their risk. The rate of return that investors require will reflect both the riskiness of the future cash flows and the anticipated rate of inflation. Since accounting earnings do not account for the time value of money, we cannot use them as reliable signals to the manager who seeks to maximize shareholders' equity.

We should not blame the accountants for these inconsistencies in value-based metrics. Accounting information is prepared in accordance with generally accepted accounting principles. These principles come from a number of sources, the most important of which are the FASB and the Securities and Exchange Commission (SEC). Accounting information must be readily verifiable if it is to facilitate audited financial statements that interested users of these reports can rely upon. The emphasis on reliability often comes at the expense of relevance, however. For example, accountants record assets on the balance sheet at their initial acquisition cost (referred to as *historical cost*) rather than at current market value. While current market value is unquestionably more relevant for valuation purposes, it is often quite subjective and thus subject to reporting bias.

Since accounting information is prepared largely for the stewardship purpose of assessing historical performance, it should not be too surprising that performance measures based solely on historical accounting information have some limitations when used in a valuation context to assess future performance. Thus, the problems we have mentioned with the use of accounting-based measures do not involve the accounting system itself or the way accounting information is prepared. Instead, the difficulty is that the information is used for purposes for which it was never intended.

In part II of the book we review the new performance metrics, which we refer to as *value-based management*. We argue that value-based management is more than simply a performance tool designed to overcome the shortcomings of pure accounting-based measures. However, many of the limitations of the accounting measures that this chapter highlights are also applicable to the measures of VBM.

Part II

The Finer Details of Value-Based Management and Corporate Social Responsibility

Part I introduced both VBM and CSR in general terms. In part II we look at VBM metrics in more detail. First we discuss free cash flow, the underlying concept of all of the VBM metrics. We then consider what is arguably the most popular of the VBM metrics today, economic value added (EVA). In addition, we also discuss CSR in more detail. Finally, we highlight several firms that blur the distinction between VBM and CSR in a way that is consistent with our definition of value(s)-based management. We find that VBM and CSR are interdependent and can provide a recipe that will help businesses achieve the ultimate goals of a VBM program.

Chapter 4

Free Cash-Flow Valuation: The Foundation of Value-Based Management

> Cash never makes us happy, but it's better to have the money burning a hole in Berkshire's pocket than resting comfortably in someone else's.
>
> —Warren Buffett, chairman of Berkshire Hathaway (shareholder letter, March 1, 1999)

All of the popular methods of value-based management share a common theoretical heritage: They are, without exception, rooted in the concept of free cash-flow valuation. Specifically, they are all built upon the underlying premise that the value of any company (or of its individual strategies and investments) is equal to the present value of the future free cash flows the entity is expected to generate.

Free cash-flow analysis became the measurement standard in the 1980s and continues to be the primary method for valuing a company or a strategic business unit. In particular we see greater use of free cash-flow models in evaluating strategic decisions; for example, in acquisitions, joint ventures, divestitures, and new-product development. In earlier decades, we heard only about earnings as the primary driver of value. Today there is increasing interest—from Warren Buffett to the U.S. Postal Service—in the relevance of free cash flows and their implications for managing a firm to create value. While there has been an ongoing debate on the use of specific value-based management techniques, free cash-flow analysis as the core valuation concept has not been questioned.

The growing attention being given to free cash flows is no doubt linked to a changing philosophy. As chapter 3 emphasizes, many people no longer view the traditional accounting measures of earnings per share or return on invested capital, for instance, as adequate benchmarks of value creation. Nowadays more investors than ever believe both that "investment accounting" (another term for free cash flows) is the correct paradigm and that it bears little relationship to the historical accounting statements. In short, many have concluded that it is free cash flows that matter most.

In this chapter we define and show how to measure free cash flows and then explain how they are converted to their present value through the use of the discounted cash-flow model. We describe and illustrate the use of the free cash-flow valuation concept and discuss the role of value drivers in management's efforts to create value. Finally, we consider investors' needs and wants with regard to information (including free cash flows) about a firm.

The Beginning for Value-Based Management: Free Cash Flows

We begin with the basic notion that firm value is the present value of a company's future free cash flows. Then we restate the problem as follows:

> Firm value is the present value of the free cash flows from existing assets and includes the present value of growth opportunities.

This revision allows us to assign value to new strategies, a process that LEK/Alcar (a consulting firm well known as an enthusiastic advocate of creating shareholder value and now goes by the name L.E.K. Consulting) calls *strategic value analysis*.[1] In chapter 5 we consider the views of Stern Stewart and Company on this subject in its restatement of the free cash-flow paradigm:

> Firm value is equal to the present value of all future "economic value added" (or what the company has popularized as EVA) plus invested capital.[2]

In theory, all of these models give identical results, but then comes the rest of the story: There is much disagreement as to which approach best helps management to create value.

What Is Free Cash Flow?

To begin with the basics, what is a free cash flow? Most managers think a lot about this topic. They forecast it, worry about it, discuss it with their bankers, and constantly search for ways to improve it, but the concept is not widely understood. Just ask someone to define cash flow, and you will get a wide variety of responses that range from the balance in the firm's checking account to some ill-defined cash that results from the firm's operations.

A firm's purpose in computing its cash flow will determine the appropriate definition and measurement of this money. That is, how we intend to use our calculation matters. As one choice, we could use the conventional accountant's format, which is called a cash-flow statement. In it, the accountant explains the cause of the reported change in a firm's cash balance from one balance sheet to the next. Such knowledge, while meaningful for some purposes, has little relevance in managing the firm to create value. Instead, we are interested in the investors' perspective as to why cash flow matters. While the computations are similar, the difference between the accountant's perspective and the investors' is not merely semantic. The investors want to know the relevant cash flow in order to determine firm value, which is exactly what we want to know as well. What is the cash flow that matters to the firm's investors? It is the cash that is free and available to provide a return on the investors' capital. Simply stated, free cash flow is the amount that is available to the firm's investors.

In addition, free cash flow is one and the same regardless of whether we view it from the firm's or the investors' point of view. There is an important equality that we must understand if we are to grasp the significance of free cash flow within a valuation context:

> The cash flow that is generated as the result of a firm's operations and its investments in assets equals the cash flow paid to-received by—the company's investors.

That is,

> Firm's free cash flow = investors' cash flow.

Let us look more closely at measuring free cash flow first from the firm's viewpoint and then from the investors' perspective.

Calculating a Firm's Free Cash Flow

A company's free cash flow is equal to its after-tax cash flow from operations less any incremental investments made in the firm's operating assets. Specifically, free cash flow is calculated as follows:

operating income
\+ depreciation and amortization
= earnings before interest, taxes, depreciation, and amortization (EBITDA)
− cash tax payments
= after-tax cash flows from operations
− the investment (increase) in net operating working capital, which equals current assets less non-interest-bearing current liabilities
− the investments in fixed assets (capital expenditures) and other assets
= free cash flow

In the preceding calculation, we add back depreciation and amortization since they do not involve a cash payment. Also, the cash tax payments are the actual taxes paid, not the amount accrued in the income statement. Notice, too, that only the non-interest-bearing debt, such as accounts payable and accrued wages, are included in computing the increase in net working capital.[3]

To illustrate how to compute free cash flow, consider Johnson & Johnson's operations in 2005, a year in which they produced $2.28 billion in free cash flow. The makeup of Johnson & Johnson's free cash flow was as follows (the numbers represent billions of dollars):

operating profit		$13.371
depreciation and amortization		2.093
EBITDA		$15.464
cash taxes		(4.003)
after-tax cash flows from operations		$11.461
nonoperating income		.450
investment in current assets	$4.074	
decrease in non-interest-bearing current liabilities	1.114	
investment in net working capital		$5.188
investment in fixed assets and other long-term assets		2.727
free cash flow		$3.996

Thus, Johnson & Johnson generated $11.461 billion from operations, plus $450 million from nonoperating activity. This amount, however, was reduced by the incremental investment in net working capital of $5.188 billion and by the $2.727 billion invested in both fixed and other long-term assets.

What happened to the $3.996 billion in cash flows that Johnson & Johnson produced? Quite simply, this amount was "free" to be paid out to the firm's investors. Determining the amount of cash received by Johnson & Johnson's investors can validate this fact.

Calculating the Investors' Cash Flows

We can compute the cash flow received by a firm's investors (or what we call the financing cash flow) as follows:

  interest payments to creditors
+ repayment of debt principal
− additional debt issued
+ dividends paid to stockholders
+ share repurchases
− additional stock issued
= investors' cash flow

Thus, the investors' cash flow is simply the net cash flow paid to or received by the firm's investors, and, if negative, it is the cash flow that the investors are investing in the firm. We already know that Johnson & Johnson had $3.996 billion in free cash flow in 2005. We would expect that this amount is equal to the cash flow received by the investors, which is confirmed as follows (in billions of dollars):

interest paid to creditors	$0.165
decrease in debt principal	(0.522)
dividend payments	4.392
stock issue	(0.039)
investors' cash flow = free cash flow	$3.996

As already suggested, the firm's free cash flow is *always* the same as the cash flows remitted to the firm's investors.

To summarize, a company's free cash flow is equal to its cash flow from operations less any additional investments in working capital and long-term assets. Furthermore, it is equal to the amount distributed to its investors—thus, the name *free* cash flow.

Interestingly, when we present this equality to executives, some do not find it as intriguing and significant as we do. A few will say, "What's the big deal? All we would have to do is change the amount or form of what is paid to the investors and thereby change the free cash flow." Although this is true, they fail to understand that the amount and the makeup of a firm's free cash flow is not the result of "playing with the numbers." Instead, free cash flow is the consequence of management policies and practices that have implications to the investors with regard to the firm's value. Consequently, a company's free cash flow is the result of operating, investing, and financing decisions and is not some ad hoc number that can be manipulated as we want.

Why Free Cash Flow Matters

Even the infamous Motley Fool, a web-based investment advisory service, recognizes that, to some investors, free cash flows drive a firm's value. The following is an excerpt from a 2005 Motley Fool article on free cash flow:

> We talk about free cash flow a great deal around here and with good reason. It is the gold standard by which to measure the profitability of a company's operations. Free cash flow is not perfect, but it is more difficult to manipulate than net income or earnings per share. For this reason, it is also likely to be lumpier than net income.
>
> Caveats aside, free cash flow is my favorite metric when evaluating a company because of the insight it gives an investor as to how heavy or light a company's business model is and how clearly it shows a company's ability to reward investors. *It's what you do with it that counts.*
>
> As great as it is to find a company with strong free cash flow, the metric itself really only lets you peer into the operational profitability of a business. What a company does with its free cash flow is just as important as having it in the first place. Companies that simply hoard their cash or spend it aimlessly on acquisitions will likely do more harm than good to your portfolio.
>
> As an investor, you're much better served to look for companies that take their free cash flow and put it toward share repurchases when their shares are below their intrinsic value or, better yet, toward a regular cash dividend. The beauty of being an income

investor and receiving a cash dividend is not only that you get a guaranteed tangible return but that you also have the option to reinvest the money received in more shares of the same business or another opportunity. The point is that you get to decide how the cash is allocated.

Just make sure that the company is like ExxonMobil, which pays out a portion of its free cash flow as a dividend, unlike ConAgra, which pays out more in dividends than it generates in free cash flow.

Foolish final words

As investors, we all want companies that generate or will eventually generate free cash flow. This is equally true for currently unprofitable high-growth stories such as Sirius and XM Satellite Radio as it is for more mature companies such as Alcoa or Wendy's. This is because all a stock price represents is the market's estimate of the future free cash flows a business will generate. Following an income-investing strategy, for example, you would prefer to choose companies that currently generate free cash flows and that either pay a dividend or are likely to pay a dividend because this is a better indicator of a solid business than a dividend funded by ample free cash flow.

Nathan Parmelee (TMF Doraemon), *The Motley Fool*, August 12, 2005

Free Cash Flow and Firm Valuation

The makeup of a firm's cash flow and the way in which it is distributed to investors is essential information in managing a firm's cash resources, particularly those of a growth firm. However, as already noted, there is another reason for computing a firm's cash flow, and that is to ascertain the company's value.

Let us return to our Johnson & Johnson example, where the firm's investors received $3.996 billion in cash flows in 2005. Assume that Johnson & Johnson's investors expected the firm to generate this same cash flow of $3.996 billion each and every year into the future. What does this suggest to us about the firm's value?

Within the context of a discounted cash-flow model of firm value, we could think of Johnson & Johnson's value as equal to the present value of its expected future cash-flow stream. In other words, its value is the present value of the future cash-flow stream of $3.996 billion, discounted at the investors' required rate of return.

Valuing the Firm: Framing the Analysis

Using free cash flow for valuing a firm is relatively straightforward. Firm value is the present value of future cash flow for the entity as a whole. Specifically, a firm's intrinsic value is equal to the present value of its free cash flows discounted at the company's cost of capital, plus the value of the firm's nonoperating assets:

$$\text{firm value} = \text{present value}\begin{pmatrix}\text{free cash}\\ \text{flows}\end{pmatrix} + (\text{value of nonoperating assets}).$$

Nonoperating assets include all assets that are not necessary for the firm's operations, such as marketable securities, excess real estate, or overfunded pension plans.

To this point, we have essentially stated the obvious: To value a firm, we project future free cash flows and then determine their present value. However, framing the analysis is a bit problematic.

Free Cash Flows, but for How Long?

Projecting a company's cash flow for the life of the firm is no easy task. We could conceivably do what one Japanese firm has reportedly done, and that is to develop a strategic plan for 250 years, estimate the expected cash flow, and then calculate the present value. However, given the difficulties with forecasting distant cash flow, a more sensible approach is to divide the firm's cash flow into two parts: (1) cash flow to be received during a finite period that corresponds to the firm's strategic planning period, and (2) cash flow to be received after the strategic planning period. For example, Texas Instruments uses its "long-range plan" to project cash flow for ten years into the future. It then computes the present value of the projected operating cash flow for the planning period and the present value of the "residual value" (the value beyond the ten-year projections) to estimate the company's value. If the value is consistent with the current market price, then, according to management, "future plans are in line with what investors expect of our financial performance" (pers. comm.).

The length of the planning period should correspond to the duration of the competitive advantage the firm enjoys. Only when a company has a competitive advantage can management expect to earn returns in excess of its cost of capital. When the competitive advantage has dissipated, there is no incentive (at least not in terms of creating economic value) to continue growing the firm. Thus, growth duration is an important criterion for determining the length of the planning period.

To identify a firm's growth duration, we have to examine the company relative to its competition according to a number of factors. Consideration should be given to the presence of established distribution channels and any brand names the firm might own, as well as the firm's R&D efforts. The pharmaceutical industry, for instance, has a relatively long growth duration period because of patented products, proven processes, and R&D investment, all of which raise the barriers of entry. In contrast, small companies in fragmented industries can have little (if any) sustainable competitive advantage and as such would have very short growth duration periods and very little (if any) economic value.

As to a method of estimating a firm's growth duration, we could make assumptions about the factors that affect a firm's free cash flow. We would hold these variables constant and then vary the length of the forecast until the present value of the cash flow (less future claims) equals the market price of the firm's shares. Interestingly, most companies in a given industry fall within a certain range, signaling the market's perceptions about a firm's value growth duration.

Forecasting Free Cash Flows

Once we have decided on an appropriate planning period, the task then is to estimate the firm's future cash flow. Doing so requires us to forecast the year-to-year sales for the planning period and an annual sales growth rate, which is assumed to be constant in perpetuity after the planning period. We then project both the firm's future cash flow from operations and the asset investments to be made over time.

In asking what is important in managing for value, we believe the answer is clear: Free cash flow—not earnings—are the key determinant of value as determined in the capital markets. However, in forecasting a firm's free cash flow, we should not completely discount the informational content of earnings. It could very well be that past earnings provide a better basis for predicting future cash flow than the actual history of cash flow itself. Although earnings measure the results of operating cycles, they involve judgment, which may induce bias.[4] Cash flow, on the other hand, involves less judgment but does not measure the results of operating cycles. Thus, reported earnings can be helpful as a starting point for forecasting free cash flow.

The process of computing free cash flow can best be explained by an example. At one time we were asked to value a regional trucking firm, Central Freight Lines (CFL). To begin, we examined the firm's historical performance; we then studied the industry in which CFL competed and its competitive position within the industry. The key issues of concern were the following:

- the sales for the most recent period
- the estimated sales growth rate for the planning period and a growth rate that could be maintained in perpetuity after the planning period (the latter typically approached the inflation rate)
- the expected operating profit margins: operating profits ÷ sales
- the projected ratio of operating assets to sales, including net working capital, fixed assets, and other long-term assets relative to sales
- the cash tax rate

These variables have come to be known as the "value drivers" because they affect a firm's free cash flows, which in turn affect its value. The term *value driver* can be used to describe any item that creates value within a company. We have more to say about other value drivers in later chapters.

Based on what we learned about CFL and its industry, we made several assumptions as a beginning point for estimating the firm's free cash flow (see table 4.1). These assumptions were based on the company's historical performance and adjusted for some anticipated changes. For instance, management had developed a strategy that it believed would allow the company to increase its sales ($240 million in the preceding year) by about 8% for five years, which would then decline to 5% the next five years and then track the industry's inflation rate of 2.6% thereafter. The sales projections were estimated by considering unit sales growth, price increases, and the developing trends within the trucking industry. Management further believed the following:

- The firm's before-tax operating profit margins would remain relatively stable at 7%.
- Net working capital and other long-term assets had closely followed sales at 5.5% and 2% of sales, respectively.
- Fixed assets had increased disproportionately to sales in the past; thus, management believed fixed assets needed to be reduced relative to sales. At the time, fixed assets were 45% of sales, but management decided to reduce any additional investments in fixed assets to 40% of incremental sales during the following five years and hopefully to 35% from that point forward.
- Other long-term assets were expected to be about 2% of sales.

Management also estimated that the firm was holding excess real estate worth $7.5 million that was not essential to the firm's operations.

Table 4.1. Assumptions for Estimating Free Cash Flows at Central Freight Lines

	Value Driver Assumptions		
	1–5 Years	6–10 Years	11 Years
Sales growth	8.0%	5.0%	2.6%
Operating profit margins	7.0%	7.0%	7.0%
Cash tax rate	27.0%	27.0%	27.0%
Net working capital/sales	5.5%	5.5%	5.5%
Fixed assets/sales	40.0%	35.0%	35.0%
Other long-term assets/sales	2.0%	2.0%	2.0%

Table 4.2 presents the results of the free cash-flow calculations (in thousands of dollars) for CFL for a ten-year planning period and also for the eleventh year. The eleventh year is the first year of the *residual period,* when a constant growth rate in sales is assumed to begin and continue in perpetuity. The following might help explain these computations:

- The annual sales for the first year are based on the beginning sales of the prior year ($240 million) plus the projected 8% annual growth rate anticipated by the planned strategy. For instance:

$$\text{sales in year 1} = [1 + \text{sales growth rate}] \times \text{prior year sales}$$
$$= [1+ 0.08] \times \$240{,}000 = \$259{,}200$$

  This logic was used to determine sales in all ensuing years.
- Before-tax operating profits were assumed to be 7% of sales for all years.
- The cash taxes were projected to equal 27% of before-tax operating profits.
- The incremental investments for the different asset categories are based on the following calculation:

$$\begin{pmatrix}\text{incremental asset} \\ \text{investment in year } t\end{pmatrix} = \left(\begin{bmatrix}\text{sales in} \\ \text{year } t\end{bmatrix} - \begin{bmatrix}\text{sales in} \\ \text{year } t-1\end{bmatrix}\right) \begin{pmatrix}\text{assets-to-sales} \\ \text{percentage}\end{pmatrix}$$

Thus, for year 1 (in thousands of dollars),

$$\begin{aligned} \text{net working capital} &= [\$259{,}200 - \$240{,}000] \times 5.5\% = \$1{,}056 \\ \text{fixed assets} &= [\$259{,}200 - \$240{,}000] \times 40\% = \$7{,}680 \\ \text{other long-term assets} &= [\$259{,}200 - \$240{,}000] \times 2\% = \$384 \end{aligned}$$

Table 4.2. Central Freight Lines: Free Cash-Flow Calculations ($000)

Years	1	2	3	4	5	6	7	8	9	10	11
Value Drivers											
Sales growth	8.0%	8.0%	8.0%	8.0%	8.0%	5.0%	5.0%	5.0%	5.0%	5.0%	2.6%
Operating profit margins	7.0%	7.0%	7.0%	7.0%	7.0%	7.0%	7.0%	7.0%	7.0%	7.0%	7.0%
Cash operating tax rate	27.0%	27.0%	27.0%	27.0%	27.0%	27.0%	27.0%	27.0%	27.0%	27.0%	27.0%
Net working capital/sales	5.5%	5.5%	5.5%	5.5%	5.5%	5.5%	5.5%	5.5%	5.5%	5.5%	5.5%
Fixed assets/sales	40.0%	40.0%	40.0%	40.0%	40.0%	35.0%	35.0%	35.0%	35.0%	35.0%	35.0%
Other assets/sales	2.0%	2.0%	2.0%	2.0%	2.0%	2.0%	2.0%	2.0%	2.0%	2.0%	2.0%
Free Cash Flow											
Sales	$259,200	$279,936	$302,331	$326,517	$352,639	$370,271	$388,784	$408,223	$428,635	$450,066	$461,768
Operating profits	$18,144	$19,596	$21,163	$22,856	$24,685	$25,919	$27,215	$28,576	$30,004	$31,505	$32,324
Taxes	4,899	5,291	5,714	6,171	6,665	6,998	7,348	7,715	8,101	8,506	8,727
After-tax operating profits	$13,245	$14,305	$15,449	$16,685	$18,020	$18,921	$19,867	$20,860	$21,903	$22,998	$23,596
Incremental Investments											
Net working capital	$1,056	$1,140	$1,232	$1,330	$1,437	$970	$1,018	$1,069	$1,123	$1,179	$644
Fixed assets	7,680	8,294	8,958	9,675	10,449	6,171	6,480	6,804	7,144	7,501	4,096
Other assets	384	415	448	484	522	353	370	389	408	429	234
Total investments	$9,120	$9,850	$10,638	$11,489	$12,408	$7,494	$7,868	$8,262	$8,675	$9,108	$4,973
Free cash flow	$4,125	$4,455	$4,812	$5,196	$5,612	$11,427	$11,999	$12,599	$13,228	$13,890	$18,623
Present values	$3,619	$3,428	$3,248	$3,077	$2,915	$5,206	$4,795	$4,417	$4,068	$3,747	

Moreover, in computing the free cash flows, we did not add back depreciation expense as we had when computing Johnson & Johnson's free cash flows. In the Johnson & Johnson example, we were computing the firm's *historical* free cash flows. For CFL, we were estimating *future* free cash flows. In looking forward in time, one commonly assumes that the depreciation expense equals the cost of replacing existing fixed assets. Depreciation is viewed as a proxy for reinvestment. Therefore, we typically do not add back any depreciation expense, nor do we show a cash outflow for replacing already existing depreciable assets. The only addition to fixed assets occurs when sales increase, thereby creating a need for additional fixed assets to support this increase.

Computing a Firm's Intrinsic Value

Having projected a firm's expected free cash-flow stream for both the planning period and the first year of the postplanning (or residual) period, we then discount these amounts to their present value to determine the company's economic or strategic value. If we assume that a firm's strategic planning period is t years, we would compute the present value of the planning-period free cash flows for years 1 through t as follows:

$$\begin{array}{c}\text{planning period}\\ \text{present value}\end{array} = \frac{\begin{array}{c}\text{free cash flow}\\ \text{in year 1}\end{array}}{(1+\text{cost of capital})^1} + \frac{\begin{array}{c}\text{free cash flow}\\ \text{in year 2}\end{array}}{(1+\text{cost of capital})^2} + \cdots + \frac{\begin{array}{c}\text{free cash flow}\\ \text{in year } t\end{array}}{(1+\text{cost of capital})^t}.$$

The value of the residual cash flows in year t (the end of the planning period), with cash flow beginning in year t+1 and growing at a constant growth rate in perpetuity, is calculated as follows:[5]

$$\begin{array}{c}\text{residual value}\\ \text{in year } t\end{array} = \frac{\text{free cash flow in year } t+1}{\text{cost of capital} - \text{growth rate}}$$

Finally, a firm's economic or strategic value is equal to the present value of the combined or total free cash flow:

$$\textit{intrinsic value} = \text{present value of all free cash flow}$$

$$= \left(\begin{array}{c}\text{present value of the}\\ \text{planning period}\\ \text{free cash flow}\end{array}\right) + \left(\begin{array}{c}\text{present value of the}\\ \text{residual period}\\ \text{free cash flow}\end{array}\right)$$

We can best illustrate these present-value computations by returning to the Central Freight Lines example. Relying on the predicted free cash flows in table 4.2, we calculated the firm's intrinsic value as follows (in millions of dollars):

present value of the cash flows for years 1–10	**$38.52**
present value of the cash flows for the residual value	44.06
intrinsic value	$82.58

The present value of the free cash flow for the first ten years, $38.52 million, is simply the sum of the individual present values for each of the ten years. In calculating the present value, we used a discount rate of 14% (our estimate of the firm's cost of capital). The present value of the cash flow for the first ten years is shown in the bottom row of table 4.2.[6]

Obtaining the present value of the residual cash flow requires two calculations:

1. Compute the residual value in year 10 based on the cash flow beginning in year 11 and continuing in perpetuity. This value is as follows (in millions of dollars):

$$\text{residual value in year 10} = \frac{\text{free cash flow in year 11}}{\text{cost of capital} - \text{growth rate}} = \frac{\$18.623}{.14 - .026}$$
$$= \$163.36 \text{ million}$$

2. Calculate today's present value of the residual cash-flow stream as follows (in millions of dollars):

$$\text{present value residual cash flows} = \frac{\text{year 10 residual value}}{(1 + \text{cost of capital})^{10}} = \frac{\$163.36}{(1 + .14)^{10}} = \$44.06$$

To continue, we added $7.5 million of real estate that was not being used in the firm's operations to the firm's economic value and thus arrived at a total firm value of $90.08 million; we then subtracted the firm's outstanding interest-bearing debt of $42 million, which yielded a shareholder value of $48.08 million:

intrinsic firm value	**$82.58**
plus: value of excess real estate	7.50
equals: firm value	$90.08
less: debt	42.00
equals: shareholder value	$48.08

Determining the Discount Rate

To this point, we have assumed that we know the right discount rate, that is, the firm's cost of capital, to be used in present value calculations; however, to be truthful, we do not. Measuring a company's cost of capital is no easy task. Because estimates are likely to be inaccurate, we may even choose to compute a range of discount rates rather than a single-point estimate of the cost. One Fortune 100 company, for instance, uses a band of 8–11% for its cost of capital. Nevertheless, we are left with no other choice but to estimate the firm's cost of capital as best we can. In doing so, some basic ideas guide our computations:

- A firm's cost of capital is an opportunity cost, not an out-of-pocket expense. As an economic concept, the cost of capital is based on the opportunity cost of the invested capital. As such, it is different from an accountant's concept of cost, which exists only if it is explicitly incurred. As far as the accountant is concerned, there is no cost for equity capital when computing a firm's income. However, for the financial economist, the cost of equity is as real as the cost of debt and represents one of the more significant expenses of doing business. In measuring the foregone return that the stockholder could earn elsewhere, the manager has to look to the capital markets to ascertain the opportunity cost as implied by the prevailing market price for the security.
- Since we measure a firm's free cash flow on an after-tax basis, the cost of capital should likewise be expressed after taxes.
- A firm's cost of capital (or more accurately its weighted cost of capital) should include the costs from all sources of capital, both debt and equity. It is tempting to think of the interest rate on the firm's debt as its cost of capital, especially when the firm is financing an investment entirely by debt. This idea is not correct. We need to remember that increasing a firm's debt level has implicit costs for the shareholders owing to the firm's increased risk. Thus, we should weight the costs of each and every source of capital by its

relative contribution to the firm's overall financing. Specifically, the cost is computed as follows:

$$\begin{aligned}\begin{matrix}\text{weighted cost}\\ \text{of capital}\end{matrix} &= \begin{pmatrix}\text{cost of}\\ \text{debt}\end{pmatrix}\left[1-\begin{matrix}\text{tax}\\ \text{rate}\end{matrix}\right]\left[\frac{\text{debt value}}{\text{firm value}}\right]\\ &+ \begin{pmatrix}\text{cost of}\\ \text{equity}\end{pmatrix}\left[\frac{\text{equity value}}{\text{firm value}}\right],\end{aligned}$$

where the debt and equity values relative to firm value are the percentages of the firm's total financing coming from debt and equity, respectively.

As we have already hinted, there are countless issues to be resolved and procedures that could be used in computing a company's cost of capital, but they lie far beyond the scope of our study.[7] For our purposes, we limit the discussion to a relatively simple presentation.

To understand more about the cost of capital calculation, let us continue our example of Central Freight Lines. At the time of the valuation, the management at CFL assumed that the firm's future financing would consist of 25% debt and 75% equity, with the equity coming from the retention of profits, sometimes called *internally generated funds*. The firm's before-tax cost of debt was 7.68%; thus, given a corporate tax rate of 27%, the firm's after-tax cost of debt was 5.61%: 7.68% × (1 – 0.27).

The firm's cost of equity was estimated using the capital asset pricing model, which holds that:

$$\begin{matrix}\text{cost of}\\ \text{equity}\end{matrix} = \begin{matrix}\text{risk-free}\\ \text{rate}\end{matrix} + \begin{pmatrix}\text{company}\\ \text{beta}\end{pmatrix} \times \begin{pmatrix}\text{market}\\ \text{risk premium}\end{pmatrix}.$$

A risk-free rate of 6% was chosen because of the going interest rate for Treasury bills, along with a market risk premium of 8%.[8] The firm's beta was believed to be 1.35. Given this information, we estimated the firm's cost of equity to be 16.8%:

$$\begin{matrix}\text{cost of}\\ \text{equity}\end{matrix} = 6\% + (1.35 \times 8\%) = 16.8\%$$

Using this information, we estimated CFL's weighted average cost of capital to be 14%:

	Percentage of Capital	After-tax Cost	Weighted Cost
Debt	25%	5.61%	1.40%
Equity	75%	16.80%	12.60%
Weighted Average Cost of Capital			14.00%

To conclude, computing a firm's weighted average cost of capital can involve a wide variety of techniques, some simple and some very complex. Moreover, from a conceptual basis we also have difficulty defending the idea of a single cost of capital for a firm as a whole, as opposed to a cost of capital for each and every investment. Here we have presented one of the simpler approaches to calculating a firm's cost of capital. Even so, it is the approach that most firms take.

The Value Drivers: Digging Deeper

The foregoing discussions demonstrate the process for estimating a firm's value based on the amount and timing of its expected free cash flow. The approach, while not without its limitations, provides an excellent way for management to think about managing for value.

If we believe that the capital markets assign a value to a firm based on the free cash flow generated—and there is good reason to believe this—then the free cash-flow valuation method helps us understand what drives firm value. Equally important, we can determine which value drivers have the greatest effect on a company's value. For instance, when we tested the sensitivity of CFL's value to the sales growth rates, value actually decreased as sales growth increased. Based on the assumed sales growth rates (see table 4.2), the present value of the firm's free cash flow was $82.6 million. On the other hand, if sales did not increase at all (zero growth across all years), the present value of the free cash flows would be $87.6 million. In other words, growing the firms would lower the potential value by $5 million. Stated another way, the present value of the growth opportunities is negative $5 million.

How could this be? The answer is simple: The company was not earning its cost of capital. As sales increased, profits increased but not enough to cover the cost of capital on the additional asset investments. Thus, given management's projections, for every $1 increase in sales, the firm's value would be lowered—not what management should be doing. Only if the profit margins were increased to 7.2% would the firm's value remain unchanged as sales increased. We can think of the 7.2% as the *threshold profit margin*—the operating profit margin where value neither increases nor decreases as sales change. If the firm earned an operating profit margin greater than 7.2%, then as sales increased, firm's value would increase. Otherwise, despite intentions to the contrary, management is destroying value by developing the company.

What we have just observed is a different way of looking at the problem. Instead of thinking of value as being equal to the present value of all future cash flow, we can solve for firm value by computing two distinct components of the cash-flow stream:

$$\text{intrinsic value of the firm} = \begin{pmatrix} \text{present value of} \\ \text{free cash flow} \\ \text{from existing assets} \end{pmatrix} + \begin{pmatrix} \text{present value of} \\ \text{free cash flow from the} \\ \text{firm's growth opportunities} \end{pmatrix}$$

For new high-potential firms, much of the company's value can lie in the second component of the preceding equation. In contrast, more mature companies derive most of their value from the first component—the present value of the free cash flow from existing assets. Furthermore, the only way for the second part of the valuation equation to be positive is for the firm to earn a return on the incremental invested capital that exceeds the firm's cost of capital. Central Freight Lines is an excellent case in point. We estimated that in the three years prior to the valuation, the company's value had been reduced $30 million despite the fact that the company was profitable. This same story has been played out in numerous other businesses as well.

We also found that Central Freight Line's value was highly sensitive to changes in the firm's operating profit margins, as one would expect from our discussion. The shareholders' equity values relative to different operating profit margins were as follows (in thousands of dollars):

	Operating Profit Margins	Equity Value	Change in Base Case Equity Value
	6.00%	$27,369	–$20,715
	6.50%	$37,731	–$10,353
Base case	7.00%	$48,084	0
	7.50%	$58,436	$10,352
	8.00%	$68,789	$20,705
	8.50%	$79,141	$31,057
	9.00%	$89,494	$41,410

We see that half of a percentage point change in the operating profit margin can raise or lower value by about $10 million, which is about 20% of the equity value given the current strategic plans.

Finally, CFL's firm value was sensitive to investments in fixed assets; the firm was fixed-assets intensive. For example, a 10% decrease in incremental fixed assets (e.g., a reduction from 40% to 36% of sales) resulted in a $5 million increase in the company's value. Thus, not only should management manage its income statement better, but it could also benefit the firm's shareholders by better managing CFL's fixed assets.

We examined the company's value sensitivity relative to changes in the other value drivers as well, but we need not continue. The point is clear: Firm value is affected by certain critical factors called value drivers. Understanding the significance of these value drivers is one of the most important things a management can do in its endeavor to create value.

As the Central Freight Line example clearly shows, value drivers constitute a direct connection between financial decisions and firm value and as such offer the best focus for managing value creation. However, while sales growth is a value driver, we also need to know what drives the value driver. In other words, management may find it tempting to feel good about the exercise of free cash-flow analysis, but doing so does not get down to the "shop floor." Thus, management must—if value-based management is to make any difference—know what is behind the drivers.

Figure 4.1 shows the efforts of one company's managers to link different layers of value drivers to business unit value. In this regard, they commented:

> We simulate required performance improvement or degradation to match current market price. We also simulate the impact of changes in performance on value. Each business identifies value drivers that improve cash flow, such as reducing cost, reducing inventories, collection procedures, increasing process yield, and improving cycle time.

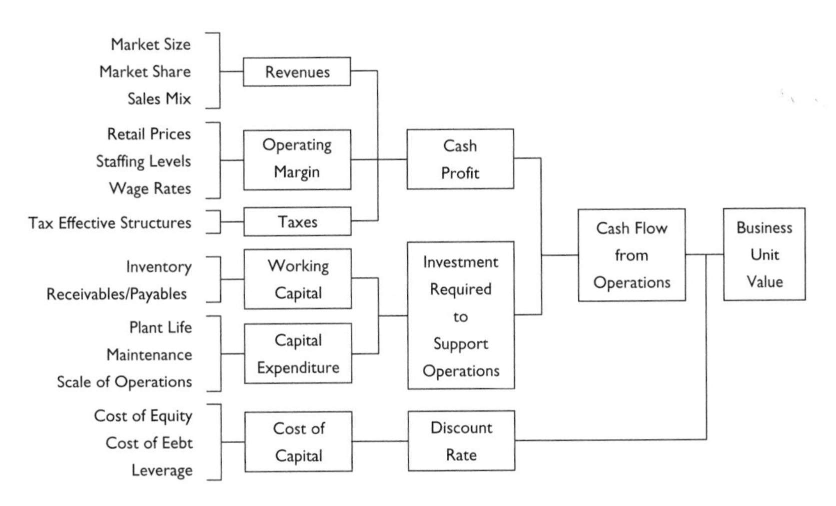

Figure 4.1. Value drivers: The key to an economic roadmap

This type of approach allows management to manage a firm for value. Only by knowing the important linkages between decisions and their effect on a firm's economic value can value-based management make a difference.

Summary

The concept of free cash flow serves as a good foundation for value-based management. No matter what we choose to do, free cash flow should be at the heart of every effort to understand how management can enhance a company's value.

Free cash flow is equal to the cash flow from operations less any incremental investments in working capital and capital expenditures. However, what makes free cash flow important is that it represents the amount distributed to the firm's investors and as such represents the core determinant of firm value.

In defining firm value, we frame the analysis as follows:

$$\text{firm value} = \text{present value}\begin{pmatrix}\text{free cash}\\ \text{flows}\end{pmatrix} + (\text{value of nonoperating assets})$$

The determinants of value (i.e., value drivers) include the following:

- amount of sales
- sales growth rates
- operating profit margins
- asset-to-sales relationships
- cash taxes

We create economic value by earning a rate of return on invested capital that exceeds a firm's weighted cost of capital. Alternatively, we can create value by not enlarging the firm (or even better, by downsizing it) when the return on capital is less than the cost of capital. Management should be committed to strategies that create value, that is, to those whose present value of the growth opportunities is positive.

Chapter 5

Pick a Name, Any Name: Economic Profit, Residual Income, or Economic Value Added

> EVA is based on something we have known for a long time: What we call profits, the money left to service equity, is not profit at all. Until a business returns a profit that is greater than its cost of capital, it operates at a loss. Never mind that it pays taxes as if it had a genuine profit. The enterprise still returns less to the economy than it devours in resources. . . . Until then it does not create wealth; it destroys it.
>
> —Peter Drucker, *The Information Executives Truly Need*

Our objective in this chapter is to give the reader a grasp of the concept of economic value added (EVA) and the calculations associated with it.[1] Of the different methods for measuring shareholder value creation, none has received more attention than EVA.[2]

While EVA can be employed for several purposes, its main use is as a period-by-period performance measurement. However, as we will see later, Stern Stewart and Co., the creator of EVA, believes it should be more than merely a financial exercise conducted within the inner chambers of a firm's financial suite.

The specific goals for this chapter are the following:

- understand the fundamental concept on which EVA is built, that is, *residual income* or *economic profits*, which are two terms that we use interchangeably
- establish the relationship between residual income (economic profits) and free cash flow
- explain the logic and rationale for EVA and the procedures for computing it
- define *market value added* and demonstrate its importance for management

The Fundamental Concept: Residual Income or Economic Profits

In chapter 3 we introduced *economic profit*, which we suggested as an alternative to the traditional return on invested capital (ROIC). We now return to this notion; however, this time we use the terms *residual income* and *economic profit* interchangeably. Moreover, economists and *managerial* accountants alike have utilized these terms. The *financial* accountant, in contrast, speaks only of *accounting* profits. Thus, the meaning of *profits* depends significantly on who is using the term. For the financial accountant, profits are measured as revenues less operating expenses less the cost of debt financing in the form of interest expense, where interest expense is the only financing cost to be recognized. There is no cost as such for the equity capital; after all, the shareholders are the owners to whom the profits flow. For economists, however, no profit is considered to be earned until the required rates of return of *all* investors, including the equity owners, have been met. In other words, true profits come only after subtracting all financing costs for both debt capital *and* equity capital, where *cost* is defined as the opportunity cost of the funds if they were invested in another firm of similar risk. In other words, those who speak of residual income maintain that a business activity must not only break even in accounting terms but also earn enough to justify the cost of all of the capital used in pursuing the activity. Thus,

$$\begin{matrix}\text{accounting} \\ \text{profits}\end{matrix} = \text{sales} - \begin{matrix}\text{cost of} \\ \text{goods sold}\end{matrix} - \begin{matrix}\text{operating} \\ \text{expenses}\end{matrix} - \begin{matrix}\text{interest} \\ \text{expenses}\end{matrix} - \text{taxes.}$$

In contrast,

$$\text{economic profits} = \text{sales} - \text{cost of goods sold} - \text{operating expenses} - \text{taxes} - \text{charge for all capital used}$$

or

$$\text{economic profits} = \text{net operating profits after taxes} - \text{charge for all capital used},$$

where the "charge for all capital used" is the after-tax interest cost on the firm's debt and a cost for its equity capital as well.

To summarize, the economic profits metric determines how well the firm has performed in terms of generating profits in a particular period—after considering the cost of the total capital that was used to generate those profits. Only if there are profits that exceed these opportunity costs would economists (and most managerial accountants) say that the firm has "made money" (i.e., earned residual income or economic profits).

Residual Income and Free Cash Flow

When it comes to shareholder value, many of us were brought up on the notion that a stock's value is equal to the present value of the future dividends (discounted dividends model). In appendix 5A we show that, given consistent assumptions, the present value of residual income and the present value of future dividends are one and the same.

The idea that value is equal to the present value of cash flow is without a doubt one of the cornerstones of finance. Thus, it is important to know how residual income relates to discounted free cash flow in determining a firm's value. Only by reconciling these two approaches can we know that value is independent of the perspective taken and that the two methods are tied to the same financial theory. In fact, they are conceptually equivalent:

$$\text{firm value} = \text{present value}\begin{pmatrix}\text{future free}\\ \text{cash flow}\end{pmatrix} = \text{invested capital} + \text{present value}\begin{pmatrix}\text{future}\\ \text{residual}\\ \text{income}\end{pmatrix}.$$

Thus, a company's value can be viewed in either of two ways: (1) as the present value of all future expected cash flow or (2) as the capital that has been invested in a company plus the present value of all future residual income. In the latter case, residual income is the value that management is creating beyond the total capital that investors have invested in the company. For instance, for a firm with a market value of $50 million (compared to $40 million of invested capital), the $10 million difference represents the market's expected future residual income stated on a present-value basis. That is, the incremental $10 million in value is the result of the investors' anticipation that the firm will earn returns greater than the cost of capital.

An Illustration of Valuation

To compare the valuation of a company in terms of free cash flows and residual income, consider the hypothetical Griggs Corporation. The company anticipates that sales in the forthcoming year (2010) will be $20 million on a total beginning capital (debt and equity) of $10 million. It also expects an after-tax operating profit margin (operating profits after taxes ÷ sales) of 6.25%, which suggests that the after-tax operating profits will be $1.25 million on the $20 million in sales ($20 million × .0625) in 2010. Given the $10 million in beginning invested capital, the firm's return on beginning-of-year capital is 12.5% ($1.25 million in operating profits ÷ $10 million of invested capital).

Assume that management is planning to reinvest 60% of its income to develop the company. Given its return on capital of 12.5%, the firm will be growing at 7.5% (60% of the 12.5% return on capital). Also, for each dollar of sales growth, $0.50 of additional investment will be required in working capital and fixed investments each year. We further assume that the 7.5% growth rate will continue for five years, which is the length of time management believes the firm can maintain its current competitive advantage period, as well as the time it can continue to earn a rate of return greater than its cost of capital of 10%. Then, after five years under the current corporate strategy, no value will be created by continued growth, Hence, there is no reason to expand the firm after the fifth year—at least not in terms of creating shareholder value.

Based on the foregoing information, we can estimate the Griggs Corporation's value either by finding the present value of the firm's free cash flow or by computing the present value of its future residual income added to the invested capital (see table 5.1). In the left-hand portion of the table, we see the free cash flow (operating profits after taxes less additional investments in working capital and fixed assets) for the five-year planning period (2010–2014), as well as its present value. In 2015 the firm's net operating profit after taxes is predicted to be $1,795,000. Since there are no plans to enlarge the firm in the year 2015 and beyond, no additional investments will be required.[3] As a result, free cash flow equals operating profits of $1,795,000, which is expected to continue in perpetuity. The value of a $1,795,000 annual perpetual cash flow stream at the end of year 2014 is $17.95 million, which is determined as follows:

$$\text{present value}_{2014} = \frac{\text{free cash flow in 2015}}{\text{cost of capital}} = \frac{\$1{,}795{,}000}{10\%} = \$17{,}950{,}000.$$

We can then discount the 2014 year-end value to today's present value (beginning of 2010):

$$\text{present value}_{2010} = \frac{\text{year 2014 value}}{(1+\text{cost of capital})^5} = \frac{\$17.95\text{ million}}{(1+.10)^5} = \$11.143\text{ million.}$$

As table 5.1 shows, the present value of all of the free cash flows—and the firm's value—is $13.314 million.

A similar process is shown in the right-hand part of table 5.1, where we find Griggs's value by taking the present value of the annual residual incomes and adding them to the beginning capital as of today. Each year's residual income is equal to the net operating profits after taxes less a charge for the beginning capital. The charge is equal to the cost of capital times the amount of beginning capital. For the year 2010, the residual income is $250,000 (i.e., $1.25 million operating profits less a 10% charge [cost of capital] on the $10 million beginning capital [$1.25 million − (0.10 × $10 million) = $250,000]). Finding the present values of the residual incomes is identical to the method used to calculate free cash flow, which results in a present value of $3.314 million for all of the future expected residual income. We then add today's invested capital of $10 million to the present value of the residual income and obtain a total firm value of $13.314 million—the exact same outcome found with the free cash flow method.

The last column of table 5.1 also shows the rate of return on invested capital (net operating profits after taxes ÷ beginning capital), which confirms that the firm is earning 12.5% on its capital (compared to a 10% cost of capital). Thus, the $3.314 million created in value comes from earning a return that exceeds the investors' 10% cost of capital and from the increasing amounts of capital used each year to expand the company. If, on the other hand, we projected only a 10% earned rate of return on the capital, the firm's value would be $10 million, which means that the present value of the residual incomes would be zero. The company's value would equal the invested capital, no more and no less.

In summary, the present value of a firm's free cash flow is the same as that of its residual income plus the capital its investors invested in the business. In theory, the two approaches do not differ, at least not when it comes to measuring a company's value. However, if a free cash flow valuation and value based on residual income are not dissimilar, then why bother? What does residual income give us that free cash flow does not?

Table 5.1. Griggs Corporation Valuation

Free Cash Flow Valuation						Residual Income Valuation			
Year	Sales	Operating Profits (after Taxes)	Investments	Free Cash Flows	Present-Value Residual Income Discounted at 10%	Beginning Capital	Residual Income	Present-Value Free Cash Flows Discounted at 10%	Return on Beginning Capital
2010	$20,000	$1,250	$750	$500	$455	$10,000	$250	$227	12.5%
2011	21,500	1,344	806	538	444	10,750	269	222	12.5%
2012	23,113	1,445	867	578	434	11,556	289	217	12.5%
2013	24,846	1,553	932	621	424	12,423	311	212	12.5%
2014	26,709	1,669	1,002	668	415	13,355	334	207	12.5%
	Present-value free cash flow (2010–2014)				$2,172				
2015	28,713	1,795	0	1,795	$11,142	14,356	359	2,229	
						Present value of all future residual incomes		$3,314	
						Original invested capital		10,000	
			Free cash flow value		$13,314	Residual income value		$13,314	

A Comparison of the Residual Income and Free Cash Flow Approaches

The weakness of free cash flow analysis is that it does not provide a readily apparent measure of *annual* operating performance. The free cash flow can be negative for one of two reasons: (1) investment is high in a profitable business, or (2) operating profitability is low in an unprofitable business. For example, in 2006 Nabors Industries LTD, a firm in the oil and gas drilling industry, had a negative free cash flow of approximately $441 million yet showed a strong return on capital of 13.5%. The negative free cash flow was the result of nearly $2 billion in capital expenditures at Nabors. In contrast, XM Satellite Radio Holdings, Inc., a satellite radio broadcasting company, had a negative free cash flow of more than $700 million, but, unlike Nabors, this was the result of large operating losses that resulted in a return on capital of –82.4%. Thus, free cash flow can be uninformative or even misleading. The old adage that "happiness is a positive cash flow" may not be as true as we have been led to believe. If all goes well, residual income will provide a better measure of period performance while maintaining consistency with free cash flow valuation. However, as we will see later, that may not turn out to be the case.

An additional advantage of using the residual income approach over a free cash flow method within a valuation context becomes apparent when we consider the inherent assumptions of each approach. The free cash flow valuation method does not attribute any value to current assets in place. Instead, all of the value is attributed to future free cash flow that is generated from these and future assets. In contrast, the residual income approach values current assets in place and then adds to this the present value of any future excess earnings, which are defined as earnings that exceed the firm's cost of capital. These different underlying assumptions for value require that the free cash flow method place far more of the value in the distant future. As anyone who has ever tried to predict the future will tell you, the farther into the future you go, the greater the uncertainty.

"Fine-Tuning" Residual Income with EVA

The acronym EVA was first used for economic value added by Patrick Finegan in 1989, but it was not until four years later that it began receiving major attention, largely the result of a feature article on the topic that appeared in *Fortune* magazine (Tully 1993). That article provided a basic presentation of the EVA concept and its computation; interviewed Joel Stern and Bennett Stewart of Stern Stewart and Co., the leading proponents of EVA use; and offered examples of major U.S. corporations that were successfully using EVA as a measure of corporate performance. From that auspicious beginning, economic value added has captured the interest of many in the business community, including firms such as Coca-Cola Co., Eli Lilly, Bausch & Lomb, Briggs & Stratton Corp., the U.S. Postal Service, and Siemens A.G. Peter Drucker has stated that EVA is a measure of "total factor productivity," whose growing popularity reflects the new demands of the information age (Drucker 2008, 345). In the words of Robert Boldt, former investment officer at CalPERS, a $240 billion pension fund, "Accounting benchmarks just don't do the job. Only EVA gives a real picture of value creation" (Tully 1998, 193). As a result, CalPERS determines its focus list of underperforming companies based on a firm's long-term stock performance, its corporate governance practices, and an economic value added calculation.

What Is EVA?

An understanding of economic value added can best begin by viewing it within the context of our earlier discussion of residual income or economic profits. Economic value added is simply a modified or, according to many EVA users, a new and improved measure of economic profits. Figure 5.1 helps us visualize the basic relationships among accounting profits, economic profits, and economic value added.

As reflected in the "earnings" line of the figure, financial accountants developed the income statement based on an accrual system for matching revenues with the relevant expenses. In the next line, we add back after-tax interest expenses to determine the firm's operating profits—the earnings available to all of the firm's investors. Next, the financial economist or managerial accountant subtracts a charge for the use of the *total* capital invested, not just for debt (as the financial accountant would do). Finally, EVA proponents make additional adjustments to the financial statements in an effort to better reflect the economic sense of the data. Generally accepted accounting principles (GAAP) do not matter to an EVA proponent if they are not considered pertinent to investors in the capital markets.

While both Joel Stern and Bennett Stewart view residual income as a definite improvement over conventional accounting profit measures, they would contend that something is still missing. They argue strongly—probably even too strongly in some accountants' view—that many of an accountant's activities are not really relevant when it comes to

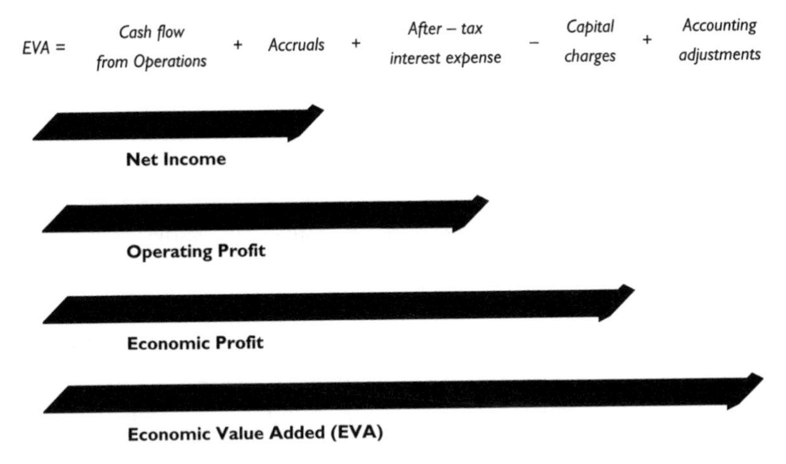

Figure 5.1. Reconciling EVA with operating cash flow, net income, operating profit, and economic profit

explaining value creation. Accountants, according to Stern and Stewart, serve some important functions, but providing the information needed to determine whether a firm is creating or destroying value is not one of them. Since investors are interested in cash flow—and not profits per se—all of the accruals and reserves the accountant creates are of questionable value. In fact, Stern and Stewart would view them as a deterrent to understanding the economic health of a firm. Thus, they believe that all of this "noise" or "distortion" needs to be removed when computing a company's economic profits and estimating the amount of capital invested by the firm's investors. Stated plainly, they would encourage analysts not to accept the accountant's measure of operating profits and asset book values at face value.

Measuring a Firm's EVA

On the surface, economic value added looks much like economic profits and is computed as follows:

$$\text{EVA} = \text{NOPAT} - (\text{k} \times \text{capital})$$

where NOPAT = the firm's operating profits after taxes but before any financing costs and noncash bookkeeping entries, except for depreciation;[4]

k = firm's weighted average cost of capital; and

capital = total cash invested in the firm over its life but net of depreciation.

Alternatively, EVA is frequently expressed as follows:

$$\text{EVA} = (\text{r} - \text{k}) \times \text{capital},$$

where r is the firm's return on capital, computed as follows:

$$\text{return on capital} = \frac{\text{NOPAT}}{\text{capital}}.$$

Based upon the foregoing measurements, we can determine whether a firm is creating or destroying value (EVA would be positive or negative). Specifically, we know that management can increase a firm's value in one of the following ways:

- increase the rate of return earned on the existing base of capital (i.e., generate more operating profits without tying up any more capital in the business)
- invest additional capital in projects that return more than the cost of obtaining the new capital
- liquidate capital from (or at least curtail further investment in) operations that are earning inadequate returns

Briggs & Stratton and EVA

Each year in its annual report Briggs & Stratton includes a section it calls "performance measurement," which shows the firm's EVA computation for the prior three years. The 2005 computations and explanations are presented below. Management subscribes to the premise that the value of Briggs & Stratton is enhanced if the capital invested in its operations yields a cash return that is greater than that expected by the providers of the capital.

Conventional financial statements and measurements, such as earnings per share and return on shareholders' investment, are of less interest to the providers of capital than indicators of cash-flow generation and effective capital management. Consequently, we adhere to a measurement of performance that guides operational and corporate management in evaluating current decisions and long-term planning strategies and steers it toward the goal of maximizing cash operating returns in excess of the cost of capital. The following table summarizes the results for fiscal years 2003–2005 (in thousands):

	2005	2004	2003
Return on Operations			
Income from operations	$190,768	$234,209	$149,922
Adjust for:			
Other income except interest income	15,084	5,490	6,545
Increase (decrease) in:			
Bad debt reserves	280	(196)	77
LIFO reserves	1,056	4,103	2,551
Warranty accrual	5,376	(4,442)	1,244
Adjusted operating profit	212,564	239,164	160,339
Cash taxes[a]	(74,886)	(69,065)	(27,833)
Extraordinary gain – net of taxes[b]	19,800	—	—
Net adjusted cash operating profit after taxes	$157,478	$170,099	$132,506
Weighted average capital employed[c]	$1,612,977	$1,278,586	$1,164,605
Economic return on capital	9.8%	13.3%	11.4%
Cost of capital[d]	8.6%	7.7%	8.4%
Economic value added	$18,761	$71,648	$34,679

a. The reported current tax provision is adjusted for the statutory tax impact of interest income and expense.

b. Represents the difference between fair market value of the assets acquired from Murray, Inc. (and Murray Canada Co.) and the cash paid, net of taxes.

c. Twelve-month weighted average of total assets less non-interest-bearing current liabilities plus the bad debt, LIFO and warranty reserves, minus deferred taxes.

d. Management's estimate of the weighted average of the minimum equity and debt returns required by the providers of capital.

Calculating NOPAT and Capital

Although computing EVA is not fundamentally different from calculating economic profits, it is the adjustments going into computing EVA that make it unique. These adjustments, or what Stern Stewart and Co. calls *equity equivalents,* are made for the express purpose of converting both NOPAT and capital from an accounting book value to an economic book value. While Stern Stewart speaks of some 160 or so possible equity equivalents for a particular firm, five to ten adjustments are more typical.

It is quite difficult to find a unifying theme for the set of potential adjustments. We have heard that these adjustments are designed to convert the accountant's accrual-based definition of earnings to a cash-based definition. This may be the case for some adjustments, such as the elimination of reserves for inventory obsolescence or warranties. Others, however, move from cash-based to accrual-based accounting, such as the adjustment to accrue and then amortize research and development costs. We have identified two major themes that characterize the vast majority of these adjustments, although any particular modification may fit into only one theme, and in some cases a revision may be made for one reason but be inconsistent with the other theme. These two themes are the following:

1. Correcting for so-called accounting distortions that are introduced by generally accepted accounting standards. These distortions are often related to the conservative nature of GAAP. The adjustments are intended to move closer to an economic, rather than an accounting, definition of performance. An example is the capitalization of market building expenditures that have been expensed in the past (converting from a liquidating perspective to a going-concern perspective [e.g., capitalizing expensed R&D]).
2. Motivating the proper managerial behavior, which includes reducing reporting discretion, and ensuring full accountability for all of the capital that has been invested. An example of the latter is the charge that applies to all equity capital rather than just debt capital. An example of the former includes the elimination of many reserve accounts.

Note that the adjustment to accrue and then amortize R&D expenditures is intended not only to move closer to an economic definition of performance but also to avoid myopic managerial behavior. The removal of certain accounting reserves, such as one for bad debt expense, does get rid of reporting discretion and therefore does away with a temptation to "manage accounting earnings." However, it also eliminates the proper matching of revenues and expenses within a fiscal period and hence moves further from economic performance.

EVA proponents often disagree as to what constitutes the proper set of adjustments for any particular firm. Stern Stewart and Co. provides guidance on this question by recommending that an adjustment be made only if it passes the following four tests:

1. The adjustment will likely have a material effect on EVA.
2. The adjustment can be influenced by management.
3. The adjustment can be readily understood.
4. The information required to calculate and track the adjustment is readily available.

In other words, Stern Stewart takes a very pragmatic approach to the calculation of EVA. Precision is not the ultimate goal. Rather, unless the benefits of the adjustments outweigh their costs and the adjustment can be understood and influenced, there is no need to make it.

There are two equivalent approaches—from a financing perspective and from an operating perspective—for calculating NOPAT and capital. Table 5.2 provides a framework for first computing NOPAT and then capital. Both perspectives yield the same answer. While we cannot include all of the possible adjustments (equity equivalents) to be made, those shown are the primary ones. Appendix 5B provides a detailed illustration of the computation of EVA using both the financing and the operating perspective.

Table 5.2. Measuring a Firm's NOPAT and Capital

Computing NOPAT	
Financing Perspective	Operating Perspective
Income available to common + interest expense after taxes + implied interest expense on noncapitalized leases after taxes – interest and other passive investment income after taxes + preferred dividend + minority interest provision	Net operating profits before taxes (NOPBT), excluding unusual losses or gains + implied interest on noncapitalized leases – cash taxes, which equal: provision for income taxes – increase in deferred tax reserve + marginal taxes saved (paid) on unusual losses (gains) + marginal taxes saved on interest expense on debt and implied interest on noncapitalized leases – marginal taxes paid on interest and other passive investment income
+ Changes in equity equivalents Increase in deferred tax reserve Increase in LIFO reserve Goodwill amortization Increase in bad debt reserve Increase in (net) cumulative expensed intangibles (e.g., R&D and product development) Unusual loss (gain) after taxes Increase in other reserves, such as for inventory obsolescence warranties	+ Changes in equity equivalents Increase in LIFO reserve Increase in bad debt reserve Goodwill amortization Increase in (net) cumulative expensed intangibles (e.g., R&D and product development) Increase in other reserves, such as for deferred income inventory obsolescence warranties deferred income
= NOPAT	= NOPAT

(Continued)

Table 5.2. *(Continued)*

Computing Capital	
Financing Perspective	Operating Perspective
Common equity + interest-bearing debt + present value of noncapitalized leases + capitalized leases − marketable securities and construction in progress + preferred stock + minority interest + Equity equivalents Deferred tax reserve LIFO reserve Bad debt reserve Cumulative goodwill amortization Unrecorded goodwill (Net) cumulative expensed intangibles (e.g., R&D and product development) Cumulative unusual loss (gain) after taxes Other reserves, such as for inventory obsolescence warranties deferred income	Total assets − marketable securities and construction in progress − non-interest-bearing current liabilities + present value of noncapitalized leases + Equity equivalents LIFO reserve Bad debt reserve Cumulative goodwill amortization Unrecorded goodwill (Net) cumulative expensed intangibles (e.g., R&D and product development) Cumulative unusual loss (gain) after taxes Other asset-contra reserves, such as for inventory obsolescence warranties
= Total Capital	= Total Capital

Headwaters: EVA and Successful Growth

Critics of EVA claim that an adherence to EVA will stifle growth within a company. One strong counterexample to this, however, is Headwaters, Incorporated. Headwaters is the largest provider of technology and chemical reagents to the coal-based synthetic fuels industry and uses its expertise to develop new opportunities in the clean coal marketplace. According to its website, Headwaters believes in "Creating value through innovative advancements in stewardship of natural resources throughout the world."

Nonetheless, things were not always rosy at Headwaters. When Kirk Benson, the company's CEO, joined the company in 1999, Headwaters was losing money. Benson implemented an EVA-based VBM program, and, as he put it in on the website:

> The survival of Headwaters was an issue, but we persevered, and through the hard work and dedication of our employees, we were able to benefit from the synfuel opportunity and create value for our stockholders. We have come a long way from those days. In fiscal 1999, Headwaters had approximately 50 employees, revenues of $6.7 million, and a net loss of $28.4 million. We ended fiscal 2006 with over 4,000 employees and generated $1.12 billion dollars in revenue and net income of over $100 million.

Under Benson, Headwaters has steered away from its sole reliance on the Section 29 synfuel business and diversified. Benson describes the growth this way:

> In order to diversify away from the risks associated with Section 29, Headwaters initiated a two-part strategy several years ago. First, we completed several acquisitions in the construction materials industry, which generated over $854 million in revenue in fiscal 2006. Second, we have also created a portfolio of new energy and specialty chemical products. We expect that several of these new products will produce revenue in 2007 and that they will contribute substantially to Headwaters' future growth. As a result of this two-part strategy, we have developed the asset infrastructure and critical mass to realize superior organic growth in the years ahead.

Headwaters website, http://www.hdwtrs.com/

From EVA to MVA

Management's ultimate goal is neither to boost the return on invested capital nor to increase a single EVA. An individual EVA does not capture the investor's perception about management's ability to generate positive EVAs in future years. After all, it is the present value of the future EVAs that determine a firm's market value. For this reason, we need an additional measure that will indicate how the markets are assessing a company's outlook for generating future EVAs. That measure is market value added (MVA).

Market value added is the difference between a company's market value and its invested capital. In other words, MVA is the premium the market awards a company over and above the money investors have put into it, based on the market's expectations of future EVAs. Earlier we indicated that the difference between a firm's market value and its capital is equal to the present value of all future residual income. Since EVA is a modified form of residual income, we can conclude that MVA is equal to the present value of all future EVAs.

Two possible scenarios can occur. The market value of the capital is either greater than the capital invested (MVA is positive) or less than the capital invested (MVA is negative). In the first scenario, investors believe that management will more than earn the firm's cost of capital. As a result, they assign a value greater than the invested capital. However, in the second scenario, investors are signaling that they do not believe the firm will satisfy their required rate of return. What we are observing is similar to a net present value (NPV) analysis for an individual project. A project's NPV is positive if the expected internal rate of return is greater than the cost of capital; otherwise, it is negative. What management should aim for is maximizing MVA, just as it works to maximize NPV on projects.

The market recognizes and impounds value creation by granting a multiple of invested capital in excess of 1.0. Some call this the *one-dollar test.* Warren Buffett described the use of the one-dollar test in a Berkshire Hathaway letter to its shareholders of March 3, 1983: "It is our job to select businesses with economic characteristics allowing each dollar of retained earnings to be translated eventually into at least a dollar of market value."

Marakon Associates, a consulting firm that concentrates on value-based performance measuring and planning, states that, in its experience, 100% of the value created for most companies is concentrated in less than 50% of the capital employed. If this is true, there remains substantial opportunity for the management of many companies to unlock value (Michael Mauboussin 1995).

In chapter 2 we showed the top and bottom five wealth creators on Stern Stewart and Co.'s annual list of companies ranked on MVA. In table 5.3 we have expanded the list to the top twenty-five wealth-creating companies in 2006 (based on 2005 data). In the table we present not only MVAs but also EVAs. Clearly there is no perfect link between a firm's MVA and its reported current-year EVA. As we have already suggested, MVA represents the market's assessment of the firm's future EVAs, as opposed to a single historical EVA as reported here. It is also interesting to notice the diversity of firms that made the list, from high-tech to discount stores, to financial firms and home products.

Table 5.3. The 2006 Stern Stewart Performance 1000 Top Twenty-Five MVA Companies

MVA Rank	Company Name	MVA	EVA	Capital	Return on Capital (%)	Cost of Capital (%)
1	General Electric Co.	$282,545	$6,709	$124,960	12.2	6.9
2	Microsoft Corp.	229,031	8,247	28,159	40.9	11.7
3	Procter & Gamble Co.	172,734	4,131	50,270	15.8	7.2
4	Exxon Mobil Corp.	167,560	28,580	229,608	18.9	6.1
5	Wal-Mart Stores	140,705	5,199	109,393	10.8	5.8
6	Johnson & Johnson	108,223	6,601	60,857	19.0	7.8
7	Citigroup Inc.	104,438	8,902	132,947	14.7	8.0
8	Google Inc.	103,764	1,279	2,898	66.5	12.5
9	Bank of America Corp.	100,586	7,186	111,526	14.5	8.0
10	UBS AG	90,969	3,911	26,561	24.4	10.1
11	Altria Group Inc.	87,627	6,890	97,865	12.6	5.4
12	Coca-Cola Co.	83,944	3,637	18,353	25.3	5.9
13	Cisco Systems Inc.	83,169	(927)	38,106	9.7	12.0
14	Genentech Inc.	77,243	898	12,387	17.5	9.6
15	Intel Corp.	70,179	3,264	34,513	23.2	13.2

(*Continued*)

Table 5.3. *(Continued)*

MVA Rank	Company Name	MVA	EVA	Capital	Return on Capital (%)	Cost of Capital (%)
16	Qualcomm Inc.	69,915	680	7,635	21.2	11.7
17	United Parcel Service Inc.	66,397	2,010	31,851	14.1	7.4
18	PepsiCo Inc.	65,024	2,323	37,050	12.4	6.0
19	Dell Inc.	64,544	2,233	506	790.0	10.3
20	Schlumberger Ltd.	63,009	604	17,357	11.8	8.2
21	Wells Fargo & Co.	60,383	3,855	47,683	16.2	7.8
22	UnitedHealth Group Inc.	58,797	2,232	23,231	18.6	6.5
23	Home Depot Inc.	58,636	3,477	39,656	18.3	8.6
24	Intl. Business Machines	58,610	(196)	71,196	10.5	10.8
25	Amgen Inc.	56,083	1,077	31,070	13.1	9.6

More than a Financial Exercise

As you read about and visit with executives who have integrated EVA into their firm's management system, you cannot help but notice the enthusiasm—even excitement—about what the use of EVA has done for the company's culture and processes. Frequently it is more an ideology and a value system than a quantitative analysis. For many it has become the paradigm through which they see the business. At a seminar hosted by Stern Stewart and Co., Bennett Stewart spoke of the "underpinnings of EVA," or what we think of as the core values that Stern Stewart believes must accompany the effective use of EVA within a firm. Some of these ideas, which we have paraphrased, include the following:

- Corporate governance should include everyone, and everyone should have a sense of being part of the creation of the firm's value.
- Economic value added is a company's profit less a required profit; it is as if we own nothing and rent everything, but it is more a measure of economic profit. It is also a measure of value added by management, as if the company has gone through an LBO.
- Economic value added is designed to provide a shared vision; it is *the* financial management system. To attain its benefit, we must use it for *everything;* otherwise, it is ineffective; it becomes too complex if used only in certain areas.
- Managers should think and be paid like owners, both viscerally and economically. When they create value, they should share in it. EVA says, "Let's share the increase in value," which is different from a bonus. Sharing value, as opposed to receiving a bonus, is what drives behavior.

Thus, to think that EVA is simply about calculating a number—as informative as that might be—misses the point. As Joel Stern is quick to say, "Anyone can compute a firm's EVA, but it's how EVA is used that makes the difference."[5] Ehrbar (1998), a former partner at Stern Stewart and Co., would tell us that simply using EVA as a benchmark of performance probably is not worth the bother. He, along with Stewart (1991), argues long and hard that, if EVA is to matter, it must not only become the financial management system within the organization but also be tied to incentive compensation from the president down to the shop floor. If it is, they suggest, then four benefits will follow:

1. EVA relies on a new and improved measurement of return on invested capital that removes all accounting journal entries that can distort the firm's economic information and mislead management and investors about the firm's financial performance.
2. EVA provides a new and improved criterion for evaluating a firm's operating and strategic decisions, including strategic planning, capital allocations, the pricing of acquisitions and divestitures, and the setting of goals.
3. Combined with the right bonus plan, EVA can instill both a sense of urgency and an owner's perspective; managers will think and act like owners because they are paid like owners.
4. An EVA system can change a corporate culture by facilitating communications and cooperation among divisions and departments. As such, it can be a key element of a firm's internal corporate governance.[6]

Thus, if applied as Stern Stewart and Co. would advocate, EVA should provide management with the right incentives to change behavior, including the way in which it utilizes capital, rather than simply serving as a tool of financial analysis. That is the primary message that Stern Stewart and Co. would want us to hear.

AT&T and VBM: The Wrong Choice

While there are many EVA success stories, failures also occur. One high-profile case is that of AT&T. In a popular 1993 *Fortune* article that sparked the business community's attention, AT&T was included in a list of the "highly regarded major firms" that were "flocking to the concept" of EVA (Tully 1993, 38–52). The article quoted William H. Kurtz, an AT&T executive, as saying, "EVA played a significant role in the firm's decision to buy McCaw Cellular." Kurtz continued, "AT&T this year will make EVA the primary measure of business units' and managers' performance." In the firm's 1992 annual report, management was also in favor of EVA:

> In 1992 we began measuring the performance of each of our units with an important new management tool called "Economic Value Added"—"EVA" for short.... EVA gives our managers a way to track the creation of shareowner value in individual AT&T units.... We have made it the centerpiece of our "value-based planning" process. And we are linking a portion of our managers' incentive compensation to performance against EVA targets for 1993.... In summary,

> our performance planning, measurement, and reward programs are now fully aligned with the interests of the shareowners.

In the ensuing two years, AT&T more than met the established EVA targets that had been set. Then, in 1995, the firm announced the spinoffs of Lucent Technologies and NCR. With the restructuring, the firm announced that the bonus plan would be:

> adjusted to provide 50% of the incentive on the EVA level of achievement and 50% based on successful accomplishment of the restructuring transition work, including the impact on PVA [people value added] and CVA [customer value added].... Because of adjustments for the NCR write-down, the 1995 EVA target was not met, and the portion of the Chairman's annual bonus which relates to this target was reduced accordingly. The 1995 results for the PVA, CVA, and restructuring transition measurements were met. (AT&T 1996 proxy)

That same year the AT&T Compensation Committee reported:

> The committee recognizes that the Company's impending restructure will render obsolete the performance criteria established for the long-term cycles 1994–96 and 1995–97. To address this transition period and the difficulty of setting long-term financial targets while the restructure is in process, the Committee has recommended and approved that the criteria for performance periods 1994–1996 and 1995–97 are deemed to have been met at the target level. (AT&T 1996 proxy)

Around 1996 the company began to abandon EVA. The Compensation Committee reported the following:

> The Company achieved its EVA target, but the Committee noted that it did so, in part, by modifying spending plans, resulting in lower average capital deployed. The Committee therefore determined that, with respect to financial performance, the additional metric of Earnings Per Share results should be considered.... The Company achieved its EVA target, but... shareholders experienced a 9% decrease in the value of their AT&T-related holdings during 1996, though the broad market rose 20%.... In 1997, the Company will re-institute a performance share program tied to three-year relative total shareholder return ("TSR") as measured against a peer group of industry competitors. (AT&T 1997 proxy)

In 1997 the company discontinued using EVA completely and chose instead to use earnings per share and the expense-to-revenue ratio as its financial performance measures.

According to some reports, AT&T abandoned EVA as a performance measure because management was unable (or unwilling) to resolve two problems that can arise when using EVA: (1) inconsistencies between EVA and shareholder value creation that can occur when using GAAP depreciation accounting, especially when there is a "lumpiness" in the firm's investments, and (2) difficulties in establishing EVA incentive targets for management. The first issue can best be explained and demonstrated within the context of an individual project evaluation rather than in the context of a company. Thus, we return to this issue in chapter 7, where we discuss VBM models when applied to project evaluation. The second concern is considered in chapter 8. However, for now suffice it to say that it is not always easy to make EVA consistent with shareholder value despite the claims that shareholder value is just a matter of "getting to cash." Some firms may be unprepared for the accounting complexity and effort required to make an EVA computation consistent with shareholder value creation.

In distinguishing firms that continue to use EVA from those that have discontinued its use, O'Byrne (2000) makes the following observation:

> It is my judgment, based on the case studies and my broader experience that companies that take a contractual approach to management compensation, by making multi-year commitments to sharing percentages and performance targets, are much more willing to invest the time and effort required to address accounting issues that must be resolved to make economic profit consistent with shareholder value. For these companies, the accounting issues have important compensation consequences. The companies that provide the basis for the examples of accounting adjustments that reconcile acquisition economic profits with shareholder value all use the Stern Stewart EVA bonus plan design with multi-year commitments to sharing percentages and expected EVA improvement. Several of them also make multi-year commitments to fixed share stock option grant guidelines. The companies that have abandoned EVA, on the other hand, take a very discretionary approach to executive compensation. (130)

Thus, O'Byrne suggests that companies should be prepared to undertake the added accounting requirements and to adopt a strong contractual approach to executive compensation if EVA is to survive and prosper within a company.

When a firm implements EVA and later discontinues it, observers may infer that, even when a company has the best of intentions, reality may not match expectations. Certainly, no initiative of any significance, such as adopting EVA, can be accomplished without unanticipated problems. A particular value-based management technique may fit some companies better than others, so that success may be either a function of the company itself or the result of mistakes made along the way in its implementation.

Summary

EVA is based on the concept of residual income. For the financial accountant, there is no cost for equity capital. However, for the financial economist and the managerial accountant, a cost is associated with the use of equity capital—the opportunity cost of these funds. After considering this cost, we have the residual income. Still, EVA is more than residual income; it is also intended to eliminate the "distortions" created by the financial accountant that make no economic sense.

Mathematically, EVA is computed as follows:

$$EVA = \begin{pmatrix} \text{net operating profits} \\ \text{after taxes (NOPAT)} \end{pmatrix} - \begin{pmatrix} \text{cost of} & \times & \textit{invested} \\ \text{capital} & & \textit{capital} \end{pmatrix}$$

where NOPAT and capital have been restated on a cash basis—or as close to it as we can reasonably get.

Firm value can be expressed in terms of EVA and in doing so will yield the same firm value as the present value of free cash flows:

In addition, EVA causes us to focus on three ways to increase value:

- Increase the rate of return earned on the existing base of capital (i.e., generate more operating profits without tying up any more capital in the business).
- Invest additional capital in projects that return more than the cost of obtaining the new capital.
- Liquidate capital from (or at least curtail further investment in) operations that are earning inadequate returns.

Some companies produce significantly positive EVAs by investing in a large number of projects with returns only modestly greater than the cost of capital. Other firms achieve excellent results by investing in a limited number of high-return projects.

Management's ultimate goal is neither to augment the return on invested capital nor to increase a single EVA. An individual EVA does not capture the investor's perception of management's ability to generate positive EVAs in future years. After all, it is the present value of the future EVAs that determines a firm's market value. For this reason, we need an additional measure that can determine how the markets are assessing a company's prospects for generating future EVAs. That measure is market value added (MVA), which is the difference between a company's market value and the invested capital. In other words, MVA is the premium the market awards a company over and above the money investors have put into it, based on the market's expectations of future EVAs.

The primary purpose of EVA is to provide an answer to the question, is management creating value for its shareholders? However, to think that EVA is simply about calculating a number—as informative as that might be—would miss the point that Stern Stewart and Co. wants everyone to understand about EVA. If EVA is used purely as a financial exercise, few of the real benefits will be realized. In fact, Stern Stewart contends that EVA offers four advantages if used properly. In short, the intent is to use EVA as a *behavioral* tool to alter capital utilization and other incentives rather than as a tool of financial analysis.

Appendix 5A

The Equivalence of the Residual Income and Discounted Dividends Valuation Approaches

To compare residual income and the present value of dividends, we rely on the concept of *clean surplus accounting*, which means that all of the gains and losses that affect a firm's book value are also included in its profits.[7] That is, the change in book value from period to period (BV_t—BV_{t-1}) is equal to profits (P_t) less the payment of any dividends (D_t). Therefore:

$$P_t = D_t + (BV_t - BV_{t-1}) \tag{5A.1}$$

Solving for dividends in year t, we get the following:

$$D_t = P_t - (BV_t - BV_{t-1}) \tag{5A.2}$$

Residual income in period t (RI_t), stated in terms of equity (not firm) value for period t, can be expressed as follows:

$$RI_t = P_t - k(BV_{t-1}), \tag{5A.3}$$

where k is the required return on equity, which we will assume to be the same for all periods.

The discounted dividend value of a firm's equity at date 0 (E_0) can be written as follows:

$$E_0 = \sum_{t=1}^{\infty} \frac{D_t}{(1+k)^t}, \tag{5A.4}$$

If we solve equation (5A.2) for P_t and substitute the answer into equation (5A.3), the results are as follows:

$$D_t = RI_t + (1 + k)BV_{t-1} - BV_t. \tag{5A.5}$$

Now, substituting equation (5A.5) for D_t in equation (5A.3), we get the following result:

$$E_0 = \frac{RI_1 + (1+k)BV_0 - BV_1}{(1+k)^1} + \frac{RI_2 + (1+k)BV_1 - BV_2}{(1+k)^2} + \frac{RI_3 + (1+k)BV_2 - BV_3}{(1+k)^3} + \ldots \tag{5A.6}$$

By combining the terms in equation (5A.6), we get the following:

$$E_0 = BV_0 + \left(\frac{RI_1}{(1+k)^1} - \frac{BV_1}{(1+k)^1}\right) + \left(\frac{RI_2}{(1+k)^2} - \frac{BV_2}{(1+k)^2} + \frac{BV_1}{(1+k)^1}\right)$$

$$+\left(\frac{\mathrm{RI}_3}{(1+k)^3} - \frac{\mathrm{BV}_3}{(1+k)^3} + \frac{\mathrm{BV}_2}{(1+k)^2}\right) + \ldots \qquad (5A.7)$$

Simplifying the equation gives us the following:

$$E_0 = BV_0 + \left(\frac{\mathrm{RI}_1}{(1+k)^1}\right) + \left(\frac{\mathrm{RI}_2}{(1+k)^2}\right) + \left(\frac{\mathrm{RI}_3}{(1+\mathrm{k})^3} - \frac{\mathrm{BV}_3}{(1+k)^3}\right) + \ldots, \qquad (5A.8)$$

and if we extend the expression to $t = \infty$ and assume that $\frac{BV_\infty}{(1+k)^\infty} \to 0$, then

$$E_0 = BV_0 + \sum_{t=1}^{\infty} \frac{RI_t}{(1+k)^t} \quad . \qquad (5A.9)$$

Thus, the dividend discount model can be restated in terms of book value and residual income (or economic profits), which is not true for traditional GAAP-based accounting profits. This restatement is possible because residual income incorporates a charge for all capital, both debt and equity, as is done in determining economic profits, P_t.

Appendix 5B

An Illustration of the Computation of EVA

To illustrate the process, we calculate the EVA for the hypothetical Hobbs-Meyer Corporation for the year 2008. The firm's financial statements are shown in table 5B.1. Equally important is the information taken from the footnotes to the financials since most of the equity equivalents are found there rather than in the statements themselves. For example, from the footnotes we learn the following:

1. The firm uses LIFO (last in, first out) for reporting its inventories; the cost of goods sold and the reserves are $175,000 and $200,000 for 2007 and 2008, respectively.
2. The company has commitments in the form of noncapitalized leases, the present values of which are $200,000 and $225,000 in each of the two years. The implied or imputed interest on these leases in 2008 is estimated at $21,000.
3. By using a pooling of interest, the firm acquired another business, which resulted in unrecorded goodwill of $40,000. However, in other acquisitions, goodwill was being recorded and amortized until 2001, when GAAP stopped requiring this annual amortization. The cumulative goodwill the firm had expensed is $73,000.

The firm's cost of capital is 10%, and its marginal tax rate is 34%.

Given the financial information we have about Hobbs-Meyer, we have calculated the firm's EVA for the year ending 2008. These computations, first for NOPAT and then for capital, are shown in table 5B.2. Both financing and operating perspectives are presented so that we can compare the two approaches:

NOPAT: Financing Perspective

We begin with the income available to the common stockholders and add back all of the financing-related expenses and interest income (e.g., interest expenses and preferred stock dividends). As part of the financing expenses, we have included an imputed interest cost associated with the noncapitalized leases. These adjustments are all restated on an after-tax basis. We then add all of the *increases* in equity equivalents (e.g., reserves, deferred income, and goodwill amortization) in order to

Table 5B.1. Hobbs-Meyer Corporation Balance Sheet ($000) and Income Statement

Balance Sheet		
	2007	2008
Cash	$16	$20
Marketable securities	4	5
Accounts receivable	$300	$350
Bad debt reserve	20	25
Net receivables	$280	$325
Inventory	2,650	3,350
Total current assets	$2,950	$3,700
Land	210	263
Plant and equipment	2,475	3,114
Gross fixed assets	$2,685	$3,377
Accumulated depreciation	(500)	(690)
Net fixed assets	$2,185	$2,687
Goodwill	50	50
Total assets	$5,185	$6,437
Accounts payable	$1,040	$1,350
Accruals	406	530
Income taxes payable	120	125
Short-term debt	110	25
Current portion long-term debt	25	27
Total current debt	$1,701	$2,057
Senior long-term debt	210	190
Capitalized leases	880	1,010
Total debt	$2,791	$3,257
Deferred income taxes	78	94
Deferred income	15	20
Preferred stock	20	25
Minority interest	25	25
Common equity	$56	$57
Paid in capital	170	175
Retained earnings	2,030	2,784
Common equity	$2,256	$3,016
Total debt and net worth	$5,185	$6,437

(Continued)

Table 5B.1. *(Continued)*

Income Statement	
	2008
Sales	$20,650
Cost of goods sold	15,900
Depreciation expense	210
Cross profit	$4,540
Selling and administrative	3,400
Operating profits	$1,140
Interest expense	135
Interest income	5
Extraordinary gains	40
Preferred dividends	3
Provision for minority interests	5
Income before taxes	$1,042
Income tax provision	488
Income available to common	$554

(1) convert from an accrual to a cash basis and to (2) remove the effects of unusual gains. The resulting NOPAT is $686,000.

NOPAT: Operating Perspective

With this approach, we begin with the *before-tax* operating profits and add the *before-tax* implied interest on noncapitalized leases. We then convert the provision for taxes in the income statement, reported on an accrual basis, to a cash basis. We also recognize any tax effects of financing costs and unusual gains. Finally, we add the *increases* in equity equivalents to convert from an accrual to a cash basis. Again, we find NOPAT to be $686,000.

Capital: Financing Perspective

To compute capital according to the financing perspective, we take the common equity investment and add all of the sources of debt, except for non-interest-bearing current liabilities (e.g., accounts payable and accrued operating expenses), preferred stock, and minority interests. We subtract any nonoperating assets—in this case, marketable securities. Finally, we add the equity equivalents—not merely the increases as we did with NOPAT but the total amounts as well. We find the capital to be $3,984,000 and $4,825,000 at year end 2007 and 2008, respectively.

Table 5B.2. Hobbs-Meyer Corporation: Computing NOPAT and Capital

NOPAT: Financing Perspective		
		2008
Income available to common		$547
Plus:		89
Interest expense after taxes		14
Implied interest on non-capitalized leases (after tax)		3
Preferred dividend		5
Minority interest provision		
Less:		
Investment income after taxes		(3)
Plus changes in equity equivalents		
Increase in deferred tax reserves		16
Increase in LIFO reserve		25
Increase in bad debt reserve		5
Increase in deferred income		5
Unusual gains (after taxes)		(26)
NOPAT		$686
NOPAT: Operating Perspective		
Operating profits		$1,140
Implied interest on noncapitalized leases		21
Net operating profits before taxes (NOPBT)		$1,161
Less cash taxes:		
Provision for income taxes		(488)
– increase in deferred tax reserve	$16	
– marginal taxes on unusual gains	14	
+ marginal taxes on interest expense	(46)	
+ marginal taxes on implied interest	(7)	
– marginal taxes paid on investment income	2	
Cash taxes		$(510)
Plus changes in equity equivalents		
Increase in LIFO reserve		$25
Increase in bad debt reserve		5
Increase in deferred income		5
NOPAT		$686

(*Continued*)

Table 5B.2. *(Continued)*

Capital: Financing Perspective		
	2007	2008
Common equity	$2,256	$3,016
Plus:		
Interest-bearing debt	345	242
Capitalized leases	880	1,010
Present value of noncapitalized leases	200	225
Preferred stock	20	25
Minority interest	25	25
Less:		
Marketable securities	(4)	(5)
Plus equity equivalents		
Deferred tax reserve	78	94
LIFO reserve	175	200
Bad debt reserve	20	25
Cumulative goodwill amortization	73	73
Unrecorded goodwill	40	40
Cumulative unusual gains after taxes	(139)	(165)
Deferred income reserve	15	20
Total equity equivalents	$262	$287
Total Capital	$3,984	$4,825

Capital: Operating Perspective		
Total assets	$5,185	$6,437
Less:		
Marketable securities	(4)	(5)
Non-interest bearing current liabilities	(1,566)	(2,005)
Plus:		
Present value of non-capitalized leases	200	225
Plus equity equivalents		
LIFO reserve	175	200
Bad debt reserve	20	25
Cumulative goodwill amortization	73	73
Unrecorded goodwill	40	40
Cumulative unusual gains after taxes	(139)	(165)
Total equity equivalents	$169	$173
Total Capital	$3,984	$4,825

Capital: Operating Perspective

This time we begin with the firm's total assets as reported on the balance sheet and subtract the marketable securities and the non-interest-bearing debt; add the present value of the noncapitalized leases; and finally add the equity equivalents related to the firm's asset accounts, such as goodwill amortization and cumulative unusual gains. It should be no surprise that the firm's capital accounts again equal $3,984,000 and $4,825,000, as found with the financing perspective.

Now that we know the NOPAT and capital for the Hobbs-Meyer Corporation, we can easily compute its EVA for 2008:

$$\text{EVA} = \text{NOPAT} - \begin{pmatrix} \text{cost of} & \times & \text{invested} \\ \text{capital} & & \text{capital} \end{pmatrix}$$

$$= \$686{,}000 - (10\% \times \$3.984{,}000) = \$288{,}000.$$

Alternatively, we can compute EVA as follows:

$$\text{EVA} = \begin{pmatrix} \text{return on} & - & \text{cost of} \\ \text{capital} & & \text{capital} \end{pmatrix} \times \text{invested capital},$$

where $\dfrac{\text{return on}}{\text{capital}} = \dfrac{\text{NOPAT}}{\text{capital}} = \dfrac{\$686{,}000}{\$3{,}984{,}000} = 17.2\%.$

Thus,

$$\text{EVA} = (17.2\% - 10\%) \times \$3{,}984{,}000 = \$288{,}000.$$

We can therefore conclude that Hobbs-Meyer created $288,000 in value for its shareholders by earning a 17.2% return on $3,984,000 beginning invested capital, compared to a weighted average cost of capital of 10%.

Appendix 5C

Performance Evaluation Using CFROI

When we think of economic value added or market value added, the name Stern Stewart comes to mind. In like manner, the term *cash flow return on investment* (CFROI) is associated with the Boston Consulting Group (BCG) and Credit Suisse First Boston Holt (CSFB Holt). For the most part BCG works with strategic planners of large, publicly held firms, while CSFB Holt has devoted itself to advising professional money managers.

While we find that EVA is still being actively promoted as a leading VBM metric, the same does not appear to be the case with CFROI. When we contacted BCG to learn more about how the company is currently using CFROI, we met resistance to our inquiries. Even though BCG claims to continue to use CFROI, we found only scant mention of this metric when we searched its website. In contrast, CSFB Holt still actively promotes the CFROI metric for investor use in portfolio evaluation and management.

Both EVA and CFROI are based on the underlying free cash-flow framework; however, they differ in one important respect. While EVA expresses value creation in terms of dollars, CFROI expresses it as a percentage. Every student of elementary finance knows about the rate-of-return analog to dollar-value performance metrics, which is generally called *internal rate of return*. Proponents of CFROI contend that percentage rates of return facilitate (1) comparisons of returns to costs of capital, (2) empirical testing of cash-flow models that employ those rates of return as primary drivers, and (3) the elimination of the size bias inherent in the use of performance metrics based on dollar values or dollar value added, which makes comparisons possible among projects, divisions, or companies.

Although the procedure for estimating a firm's rate of return can become a bit tedious at times, the concept is quite simple. Much like the interest rate earned on a savings account, we want to know the cash-on-cash return being earned. Investors make cash investments and expect to receive cash flows in return. Whether they are satisfied with the investment depends on the rate of return earned on it, compared to some minimum acceptable return.

Typically, CFROI computations begin by converting income statement and balance sheet information into cash-flow return on investment, which its proponents claim more closely approximates a firm's underlying economics. As the CFSB Holt website states, this procedure simply "takes accounting information, converts it to cash and then values that cash." The procedure illustrated on the website then describes the CFROI calculations as follows:

> The CFROI metric result is calculated in two steps: First, we measure the inflation-adjusted gross cash flows available to all capital owners in the company and compare that to the inflation-adjusted gross investment made by the capital owners. We then translate this ratio of gross cash flow to gross investment into an Internal Rate of Return (IRR) by recognizing the finite economic life of depreciating assets and the residual value of non-depreciating assets. The CFROI result approximates the economic return produced by the firm's projects.

Chapter 6

Corporate Social Responsibility: Putting the *S* in Value(s)-Based Management

> If corporations were to analyze their prospects for social responsibility using the same framework that guide their core business choices, they would discover that CSR can be much more than a cost, a constraint, or a charitable deed—it can be a source of opportunity, innovation, and competitive advantage.
>
> —Michael Porter and Mark Kramer, "Strategy and Society: The Link between Competitive Advantage and Corporate Social Responsibility"

Chapter 5 introduced the VBM metric EVA and explained how it can be used for project evaluation and incentive compensation, two very important functions of a VBM program. However, the metric simply tells the score; it does not explain *how* to score. We cannot stress this enough. We feel that many VBM failures occur when an organization focuses too narrowly on the metric and pays insufficient attention to its value drivers. Shareholder value is not created by EVA, CFROI, or any of the other VBM metrics. Rather, when a company has satisfied customers who are being served by dedicated employees, a cash flow is produced that ultimately creates wealth. This cash flow then creates the value that EVA measures. It is easy to give credit to a certain metric for a firm's success (or to blame it for the firm's failure), but this misses the point entirely.

In this chapter we discuss in more detail the current movement toward corporate social responsibility (CSR) and demonstrate that being a responsible corporate citizen makes good business sense because it guides the creation of shareholder value by taking into consideration the long-term sustainability of a firm's actions. To illustrate these points we provide case studies that include VBM success stories, as well as a VBM failure. Often one can learn as much (and perhaps even more) from studying failures as from studying successes.

The Moral Argument for CSR

Often portrayed as an alternative to VBM, CSR is somehow misunderstood to be a moral movement without a business purpose. Perhaps the following moral argument has become associated with CSR and led to this confusion:

> The purpose of a business...is not to make a profit, full stop. It is to make a profit so that the business can do something more or better. That "something" becomes the real justification for the business....It is a moral issue. To mistake the means for the end is to be turned in on oneself, which Saint Augustine called one of the greatest sins....It is salutary to ask about any organization, "If it did not exist, would we invent it?" "Only if it could do something better or more useful than anyone else" would have to be the answer, and profit would be the means to that larger end.
>
> Charles Handy (2002)

Handy's argument follows that of CSR proponents, who believe the purpose of business is too narrowly defined as simply "to make money." They believe instead that businesses have an obligation to their customers, employees, the communities in which they are located, and society in general that is far greater than simply creating profits for shareholders.

The Economic Argument for CSR

The economic argument for CSR is, like most economic propositions, based on economic self-interest. However, this argument does not dispute the claim that corporations have an obligation to their customers, employees, and other stakeholders. Far from it. The difference, if one exists, is more a matter of philosophical underpinnings rather than procedure. Handy proposes a moral foundation for CSR, whereas we do not voice an opinion regarding moral considerations. Rather, we believe that it simply makes good business sense to have such a program since CSR is really a holistic approach to running an organization. Jensen (2001, 9) asks, "Can corporate managers succeed by simply holding up value maximization as the goal and ignoring their stakeholders? The answer is an emphatic no. In order to maximize value, corporate managers must not only satisfy, but enlist the support of, all corporate stakeholders—customers, employees, managers, suppliers, local communities."

For the following reasons we believe that CSR represents a point of differentiation that can provide a competitive market advantage:

Human resources: CSR can be helpful with regard to recruitment and retention since potential recruits are increasingly asking about CSR policy. It can also help generate an atmosphere of pride within the organization.

Risk management: Reputations often take decades to build up but can be ruined in days by scandals and accidents. These events often draw bad publicity and attention from regulators and the courts. A strong CSR program can help prevent such episodes or can mitigate damages should one occur. Firms with existing reputation problems in their core business can engage in high-profile CSR activities in an effort to divert attention from these difficulties. Examples include British American Tobacco's health initiatives, British Petroleum's installation of alternate-energy wind turbines, and Wal-Mart's decisions to sell organic groceries and build eco-friendly stores.

Brand differentiation: Firms are constantly looking for ways to differentiate themselves from competitors in an effort to capture the consumer's dollars. Examples of organizations that have successfully used CSR as a brand differentiator include Whole Foods Market, Ben & Jerry's, and The Body Shop.

Avoidance of government interference: Organizations that make an effort to be good citizens may be better able to avoid excessive regulatory intervention. A counterexample, however, is Wal-Mart, which encounters difficulties when it attempts to enter new markets. Companies such as Target or Costco, with their superior CSR reputations, rarely face the same level of resistance.

In 2006 C. W. Goodyear, CEO of BHP Billiton, a leading natural resources company, described this phenomenon:

> BHP Billiton realized a long time ago that working in partnership with communities is more than about being a good corporate citizen. It's a powerful competitive differentiator. It has the potential

to establish us as the company of choice, giving us better access to markets, natural resources and the best and brightest employees. By doing so, we can maximize profits for our shareholders while also ensuring we do the right thing by those who are impacted by our business.

A Public Relations Nightmare

In 2003 a major controversy erupted over the decision by the Augusta National Golf Club (home of the U.S. Masters golf tournament) to not allow women as members. Although the National Council of Women's Organizations followed with a campaign aimed at the golf tournament's sponsors, the club's management refused to modify its stance, which led to even more negative publicity. Things went from bad to worse when the Ku Klux Klan announced it would attend the tournament in a show of support of the club's right to exclude women. A public relations nightmare thus resulted from the club's refusal to adopt a more enlightened stance before events got carried away.

In contrast, Bank of America, in its 2003 sponsorship of the PGA Tour event known as the Colonial, invited Annika Sorenstam, the top women's player, to compete. This invitation raised public interest in the tournament and provided Bank of America with wide praise and positive press coverage for its progressive attitude.

Bank of America is hardly alone in realizing the beneficial effects of a positive CSR program. For instance, Nike learned from its past miscues and installed a vice president for corporate responsibility and now publishes an annual CSR report. In so doing, the company has established itself as a committed corporate citizen and no longer faces near the level of negative publicity and consumer boycotts it faced a decade ago.

CSR within a VBM Framework: The Academic Evidence

While CSR does make economic sense, it is not by itself a complete theory. As Jensen (2001) argues, stakeholder theory should not be viewed as a legitimate contender for value maximization since it fails to provide managers with a means to balance the conflicting demands of the corporations' various stakeholders. Customers want high-quality goods and great services at low prices; employees want high compensation and a stress-free work environment; communities want significant social contributions; the government wants sizeable tax receipts; and so on. However, CSR leaves the manager without the decision criteria to make these necessary tradeoffs. In contrast, value maximization requires managers to provide as many resources as each stakeholder demands as long as the benefits received (i.e., long-term value created) exceed the additional costs.

Therefore, CSR makes sense within a VBM framework, what we call value(s)-based management. It is instructive to see whether this statement is supported by the existing evidence. Using a multitude of datasets, time periods, and methodologies, a plethora of academic studies have examined the way in which CSR is linked to a firm's financial performance. At first glance it is very difficult (and perhaps impossible) to draw any conclusion since the literature on this subject is inconsistent. A team of researchers (Orlitzky, Schmidt, and Rynes 2003) have attacked this problem head on by performing a meta-analysis of fifty-two studies that investigated the relationship between corporate social performance and corporate financial performance. Through their use of a rigorous methodology they have been able to find commonalities in the research. Their findings suggest that corporate social responsibility is likely to pay off in improved financial performance. Specifically, they find support for the following hypotheses: (1) Corporate social responsibility and financial performance are positively related in a variety of industry and study contexts; (2) a bidirectional causality exists between corporate social performance and financial performance, the so-called *virtuous cycle* (i.e., financially successful companies spend more on CSR outlays because they can afford to, but CSR in turn helps them to be more successful financially); and (3) corporate social responsibility is positively correlated with financial performance more because it helps firms build a positive reputation in the eyes of external stakeholders than because it contributes to internal organizational efficiency.

Recent theoretical studies have also looked at whether firms engage in CSR as a strategic tool, what some call *profit-maximizing CSR*. Baron (2001), McWilliams and Siegel (2000), and Bagnoli and Watts (2003) have each found that the benefits of CSR can outweigh its added costs by providing benefits such as reputation enhancement, the capability to charge a premium price for its output, and the ability to recruit and retain a high-quality workforce. Siegel and Vitaliano (2006) have utilized this theoretical work in their empirical study, which finds observed patterns of investment in CSR consistent with its strategic uses.

These results are entirely consistent with a survey of five hundred business executives conducted by Grant Thornton LLP in 2007. Company executives overwhelmingly believe that corporate responsibility programs can have a positive impact on their firms' financial results. "Corporate responsibility programs have moved out of the realm of public relations to become real tools for improving the bottom line," according to Grant Thornton partner Jim Mauer: "Companies are realizing that strong investment in corporate responsibility programs is both a civic obligation and a successful business strategy." The survey reports that the greatest benefit of these programs is an improvement in public opinion, customer relations, and a firm's capacity to attract and retain talent (Institute of Management Accountants 2007).

Driving the Value Drivers

In chapter 4 we discussed value drivers such as revenue growth and operating margins as the underlying sources of value creation. In this section we pursue this concept by looking at how some companies have put into practice values-based management by embracing CSR programs to boost these value drivers. In other words, these companies have discovered that CSR can yield a competitive advantage by operating in a socially responsible manner that creates long-term sustainable value for the firm and its shareholders while at the same time building value for the firm's other key stakeholders, the very essence of value(s)-based management.

As Peter Drucker has noted, *"There is only one valid definition of business purpose: to create a customer. The customer is the foundation of a business and keeps it in existence"* (Drucker and Maciariello 2004, 80). Thus, the question becomes a matter of what the business needs to do to create that customer.

Red Mountain Retail Group: Creating Value through Relationships

Red Mountain Retail Group (RMRG), a company that we have studied closely, believes it has found the answer to creating extraordinary amounts of value through extraordinary relationships. On the surface, RMRG is a firm that develops and redevelops retail and mixed-use real estate. The company believes it is much more than simply a real-estate development firm and is in fact, as its website states, "an amazing team of individuals who strive on a daily basis to make a difference in our communities; to explore our own personal limitations; and to not settle for anything less than extraordinary." The website goes on:

> RMRG was founded by an individual with a passion and vision to create an environment of caring and belonging. A unique place

> where a team could come together to explore and break through personal limitations while positively impacting communities and lives.
>
> Through these shared accomplishments and combined efforts, extraordinary relationships are developed both internally and externally—to the point that our business is driven and solely based upon relationships. You could say that relationships are our core business.

In his bestseller *Good to Great* (2001), Jim Collins states that a key attribute of great firms is their commitment to their people. Collins's "first who...then what" philosophy refers to getting the right people together first and then deciding what they should do. Red Mountain Retail Group, a developer of retail and mixed-use real estate, personifies this philosophy perhaps better than any other company.

As part of our process of getting to know RMRG, one of the authors of this book was invited to spend three days at a five-star resort participating in a company team-building exercise. This was not a charade in which only the top management team gets together for a few hours in the morning and then hits the links for the rest of the day. This event involved every single member of Red Mountain Retail Group from the CEO, Michael Mugel, to the receptionist. The event was part of a $15,000-per-year per-employee commitment to staff development. Why a five-star resort? Mugel explains, "We ask them to give freely of themselves at these exercises; we want to give back."

What occurred was much more than the typical motivational speaker telling the participants how to improve themselves. The team-building exercise was led by trainers from Productive Learning & Leisure, LLC, who are well schooled in breaking down communication barriers. The participants bared their souls regarding any relationship problems they believed were inhibiting peak performance. Michael Mugel is convinced that if you unlock your employees' potential, there is no limit to what they can accomplish. He is far less concerned about the how-to-do-it mechanics and far more interested in building trust in his workforce, as well as between his staff members and RMRG's suppliers, vendors, and, importantly, customers.

Some of the blunt exchanges observed could just as well have taken place in a therapist's office as in a corporate setting. What was accomplished, however, was every bit as impressive. A long, simmering dispute between two employees that was tearing apart two entire departments (both critical to the company's mission) was exposed for all to see. The result was a commitment from not only these former combatants but also from both departments to bury the hatchet. Doing so will enable them to cover each other's back when facing any future crisis that one of their many projects will likely throw at them.

This level of honesty (which they refer to as intimacy) may not appeal to everyone, but then again Red Mountain Retail Group is not a company like any other. When a new employee was asked how he liked working there, he replied that he has never felt so uncomfortable at any other place he has worked as he has at times at RMRG. When asked whether that was a good thing, he smiled and replied that it certainly was for his personal development and for his family and friends. In this way RMRG strives to make a difference in the lives of their employees and their employees' families, as well as in the communities they serve while simultaneously maintaining fiscal responsibility.

Red Mountain Retail Group is no stranger to creating wealth for its investors. as Mike Mugel points out, RMRG has been able to make "obscene" profits by doing things with its employees that others say simply cannot be done. Creating wealth is certainly a byproduct of RMRG's mission, but rather than focusing on the ends, it focuses on the means. Red Mountain Retail Group believes that its purpose is to unlock its employees' potential so that they can do extraordinary things for the company's customers. This strategy has allowed it to increase the market value of its properties more than twentyfold—from a little more than $50 million in 2000 to more than $1.2 billion in 2007. Distributions to its employees, who are allowed to share in the value creation, have increased from less than $15,000 in 2003 to more than $500,000 in 2007.

Southwest Airlines: The Employee Comes First

The last few years have not been kind to investors in the airline industry, which has sustained huge losses, while several airlines have sought protection in bankruptcy. Employees in the industry have also been hard hit because airlines have been cutting expenses by firing large numbers of workers and reducing the wages of those remaining. A noteworthy exception has been Southwest Airlines, the pioneer of low-cost air travel. As of 2004 Southwest's equity market capitalization exceeded the combined equity market value of all of its competitors. Its charismatic founder, Herb Kelleher, believes that the path to success lies in developing a reputation for exceptional customer service delivered by highly motivated employees, a concept that is somewhat at odds with the norm in the industry. Southwest has been able to create incredible wealth by focusing on the drivers of employee and customer loyalty. In an interview, Kelleher explained that putting employees first creates a winning strategy for every stakeholder:

> When I started out, business school professors liked to pose a conundrum: Which do you put first, your employees, your customers, or your shareholders? As if that were an unanswerable question. My answer was very easy: You put your employees first. If you truly treat your employees that way, they will treat your customers well, your customers will come back, and that's what makes your shareholders happy. So there is no constituency at war with any other constituency. Ultimately, it's shareholder value that you're producing.
>
> ...We basically said to our people, there are three things that we're interested in. The lowest costs in the industry—that can't hurt you, having the lowest costs. The best customer service—that's a very important element of value. We said beyond that we're interested in intangibles—a spiritual infusion—because they are the hardest things for your competitors to replicate. The tangible things your competitors can go out and buy. But they can't buy your spirit. So it's the most powerful thing of all.
>
> Herb Kelleher, quoted in Lucier (2004)

Herman Miller, Inc.: Creating Wealth through Design and Innovation

Herman Miller is probably best known for its award-winning office furniture, such as the Aeron chair. Herman Miller's corporate philosophy firmly ingrains social responsibility into the very fabric of its value creation, as evidenced by the numerous awards it receives annually. In 2006 alone, Herman Miller was the recipient of an environmental stewardship award, was named one of the fifty best manufacturing companies, moved up to number fourteen on *CRO* magazine's annual listing of the one hundred best corporate citizens, was named one of the top ten corporations for supplier diversity, was ranked among the most-admired companies in *Fortune* magazine's annual survey, and was selected as the sole representative of the contract furniture industry in the prestigious KLD Domini 400 Social Index for social responsibility.

Herman Miller realizes that the value it creates comes from the furniture it sells to its customers. Further, as the company explains on its website, "What arrives on the truck is furniture. What went into the truck was an amalgam of what we believe in: innovation, design, operational excellence, smart application of technology, and social responsibility."

Herman Miller shares a common belief with Red Mountain Retail Group that valuing one's employees is a formula for wealth creation. In the words of Brian Walker, president and CEO of Herman Miller,

"We value the whole person and everything that each of us has to offer, obvious and not so obvious. I believe that every person should have the chance to realize his or her potential, regardless of color, gender, age, sexual orientation, educational background, family status, skill level—the list goes on and on." Both Herman Miller and RMRG share beliefs about relationships with customers, designers, dealers, suppliers, contractors, and their own employees.

Value creation is no stranger to Herman Miller, which successfully adopted EVA in the mid-1990s. The following describes its EVA program:

> To help us make all the short- and long-term decisions that affect our company and help it to grow, we use a highly respected performance indicator, measurement, and compensation system called "Economic Value Added" (EVA), popularized by the management consultants of Stern, Stewart and Company.
>
> EVA is an internal measurement of operating and financial performance that is linked to incentive compensation for all employee-owners. Under the terms of the EVA plan, we shifted our focus from budget performance to long-term continuous improvements and the creation of economic value.
>
> When we make plans for improvements around here, we include an EVA analysis. When we make decisions to add or cut programs, we look at the impact on EVA. Every month we study our performance in terms of EVA, and this measurement system is one of the first things new recruits to the company learn. EVA has proven to be a strong corollary to shareholder value since its adoption.
>
> Herman Miller website, http://www.hermanmiler.com/CDA/SSA/Category/0,,a7-c1162,00.html

Whole Foods Market, Inc.: Creating Wealth through Whole Foods, Whole People, and Whole Planet

Perhaps the best example of a company that has embraced value(s)-based management and acknowledges that CSR can be a competitive advantage in its quest for value creation is Whole Foods Market. The company's philosophy is built around what it refers to as a Declaration of Interdependence, which has as its foundation the motto "Whole Foods, Whole People, Whole Planet." Success is measured, as the company states, "by customer satisfaction, Team Member excellence and happiness, return on capital investment, improvement in the state of the environment, and local and larger community support." The company strives to instill a clear sense of interdependence among its various stakeholders, whom it defines as the people who are interested in and benefit from the company's success.

The Whole Foods Declaration of Interdependence articulates the extent to which all of the company's stakeholders, including the shareholders, are mutually dependent. The company maintains that its customers, in accordance with Peter Drucker's views, are its most important stakeholders. Therefore, Whole Foods endeavors to go to extraordinary lengths to satisfy and delight its customers and recognizes that this cannot be done without outstanding customer service from its team members (employees). The company also realizes that its customers demand the highest-quality natural and organic products available, which requires the help of its trade partners (suppliers). Whole Foods supports the local communities its serves and acknowledges that, without their support, the company would not be in business. Whole Foods donates 5% of its after-tax profits to not-for-profit organizations. Similarly, it acknowledges the need for environmental stewardship and therefore supports sustainable agriculture, the reduction of waste, and decreased consumption of nonrenewable resources.

Finally, Whole Foods strives to create wealth. As its Declaration of Interdependence states, "We earn profits every day through voluntary exchange with our customers. We know that profits are essential to create capital for growth, job security and overall financial success. Profits are the 'savings' every business needs in order to change and evolve to meet the future. They are the 'seed corn' for next year's crop. We are the stewards of our shareholder's investments, and we are committed to increasing long-term shareholder value." Whole Foods uses EVA to evaluate its business decisions and to determine incentive compensation. The company claims that "EVA is the best financial framework that team members can use to help make decisions that create sustainable shareholder value."

Sony Blames Economic Value Added Mentality for Its Woes

At Sony, the financial picture is not nearly as sharp as the high-definition pictures its televisions display. In fiscal year 2006–2007, operating profits were only about 40% of those of the previous year. This decline should not be surprising to anymore who reads the blogs that are filled with dissatisfied Sony customers. Complaints against Sony range from lack of innovation to poor quality, poor customer service, and a complete disregard of customers.

Sony apparently has a different explanation for its current mess: The company blames EVA for many of its present problems. Although Sony was a poster child for EVA not very long ago, the company has since abandoned the metric. The company's problems stand in sharp contrast to those of a major competitor, Samsung Electronics, whose president and CFO, Doh-Seok Choi, states that every investment is "measured against its ability to create economic value added."[1]

Sony has publicly provided few details as to why it abandoned EVA; therefore, we are left to conjecture about what went wrong. It appears that EVA may indeed have been partly to blame but only because Sony failed to adopt a value(s)-based mindset instead focusing too much on the metric. We understand that Sony believed EVA caused too much of a focus on short-term measures rather than long-term value. This, unfortunately, is a situation consistent with emphasizing the metric rather than the drivers of the metric's success. In Sony's case, innovation and quality are the hallmarks of past achievement.

The following entry in the online Techdirt blog from late 2006 is similar in content to many we discovered from a quick Internet search on Sony:

> Oh, Look: Another Sony Quality Problem
> **from the *walkman-with-broken-legs* dept**
> Despite all its business troubles Sony's products have managed to maintain a strong reputation. Despite sometimes carrying a premium price over other brands, many consumers felt that Sony made solid, reliable products, and they were content to pay for that quality. But a recent string of problematic products from the company makes you wonder if something is slipping at Sony: the Play Station 3 was plagued by manufacturing delays, there was that little problem of all those Sony-made batteries getting recalled, and now it's had to announce a recall of several digital camera models after a problem first reported a year ago was found to be more widespread than first thought. Alongside the company's long-running rootkit fiasco, that's a lot of bad PR for the company to deal with, and it seems unlikely for Sony to come out unscathed. The company really is struggling to turn itself around, and the last thing it needs at this point is for its brand to take a hit and consumers to see its products as unreliable. So a note to Howard Stringer: while sorting out the company's content and technology mess is important, don't let it come at the

> expense of product quality. Sony can't afford to fail at either one at this point.[2]

The following is but a small sampling of remarks from the hordes of dissatisfied customers and former customers of Sony we found while browsing the Internet:

> The last thing Sony did right was the Walkman and I'm talking cassette not CD. The last time I remember Sony being a great product was in the mid to late 80s. Throughout the 90s and on Sony has never disappointed me by not disappointing me. They rarely don't stink (sorry for all the double negatives). And on top of that their memory, tapes, etc. are always proprietary. Sony is lame. This only reconfirms that I will NEVER buy another Sony product if I don't have to.
>
> I used to have my Sony DV videocam and my Sony digital camera plugged into my Sony Vaio desktop using proprietary Sony software to edit and share content while listening to my Sony mini-disc player.
>
> Fast forward five years. I still have a desktop, a laptop, a video camera, a music player, and a still camera, but not a single one of them is from Sony and I no longer need proprietary software for them all to easily share content.
>
> All the Sony replacements were a.) cheaper b.) use standard storage c.) use standard formats d.) easier to operate and interconnect out of the box e.) have never broken.
>
> Why would anyone continue to pay a premium for a Sony product? Even at the same price, I would choose a competitor.
>
> For those who say "can't bash a company for one bad product," I'm sure as you kept reading, you noticed that this isn't a "just one product" problem.
>
> When I speak, I speak for my entire family: parents, uncles, aunts, cousins, even friends...(Think about it—that's a lot of Sony products over a range of products lines), and all of them have confessed that no matter what the product (TV, computer, Walkman...) with the exception of Sony Music as a production company for artists (can't screw up a CD too bad, can ya?), it has broken down/gone wrong/stopped working and usually just after the warranty expires. It's usually a build-quality issue (i.e., not a consumer-at-fault thing) or part that will be expensive to replace.

> Customer service? Plenty of websites and blogs speak of the third ring of hell that is Sony CS, so I won't get into it here.
>
> Sorry, did I say no CD problems?...I completely forgot about that rootkit thing and their music CDs. How silly of me. Disregard previous comment.
>
> Sony has lost another one.

These comments go to the root of Sony's problems. To succeed, Sony needs to look at the drivers of EVA and not merely look at the EVA score. Innovation and quality are the foundation that the Sony brand was built upon, and without revving up these drivers, EVA is bound to suffer. The rootkit scandal involved a scheme that secretly installed a copy-protection program, referred to as a rootkit, on computers. The program was present on Sony music CDs. This software tool was run without the user's knowledge or consent. Moreover, when it was loaded on a computer with a CD, a hacker could gain and maintain access to the user's system.

When the considerable outcry led to numerous lawsuits, Sony's initial response demonstrated what some referred to as the company's disdain for its customers. Thomas Hesse, president of Sony BMG Music Entertainment, a global digital business, stated during a National Public Radio interview that "Most people don't even know what a rootkit is, so why should they care about it?"

Summary

This chapter has presented many economic arguments that support the idea that, because CSR appears to make good business sense, it fits in well within a VBM framework. This follows from the premise that it is important for a firm to operate in a socially responsible manner—one that considers the importance of all of its stakeholders and the way in which they contribute to the company's long-term, sustainable value creation. Economic arguments that support the business purpose of CSR include its usefulness in recruiting and retaining employees, providing reputational risk management, and helping to differentiate firm branding, as well as its utility in helping to avoid governmental scrutiny and interference. The existing academic evidence is consistent with the economic arguments in favor of developing a CSR program.

The chapter has also looked at several companies that engage in active CSR programs as a means of creating value. Some common themes are vividly illustrated by RMRG, Southwest Airlines, Herman Miller, and

Whole Foods Market. While all of these firms have metrics that can indicate how successful they have been, they focus on the drivers of success rather than the metric itself. In addition, these businesses have employed CSR programs to drive their value creation. First, each company realizes that the customer determines its ability to create wealth and ultimately maintain its very existence. Further, each one realizes that its employees (or team members) determine customer satisfaction. They have embraced corporate social responsibility as an integral part of their complete V_sBM program. Rather than narrowly focusing only on the shareholder, an enlightened approach realizes the interdependencies of the stakeholder groups and works to increase wealth by making the pie bigger for everyone, the ultimate win-win situation. One firm, Sony, has abandoned its use of a VBM metric. It appears to us that a failure to follow CSR processes—and not the company's use of a VBM metric—is largely responsible for Sony's recent troubles.

Part III

VBM Applications

In part III we consider two fundamental components of a complete VBM program, project evaluation and incentive compensation. An important feature of a proper VBM metric is the ability to provide a consistent measurement tool that can help with both project selection and project evaluation. In other words, it allows follow-up evaluation to be based on the same metric that was initially used to justify the project, thereby providing a significant improvement over traditional methods that use different measures for these two processes. Typically some form of discounted cash-flow analysis is used for project selection, and different measures such as return on invested capital or operating earnings assist in the follow-up evaluation of company performance. The use of a single VBM metric for both project evaluation and incentive compensation is the cornerstone of the entire VBM program.

Chapter 7

Project Evaluation Using the New Metrics

> What is a good investment opportunity? The answer, very simply, is any investment that is more valuable than the cost of making the investment.
>
> —Sheridan Titman, Arthur Keown, and John Martin, *Principles of Finance*

For many companies the new VBM metrics have successfully replaced the standard accounting-based tools (e.g., earnings per share, return on invested capital, earnings growth) to measure the performance of their ongoing operations. In an effort to maintain consistency with the tools they use to measure the anticipated performance of new projects, they are replacing the standard, discounted cash-flow (DCF) tools of project evaluation (net present value and internal rate of return) with the new metrics. Furthermore, we note in chapter 5 that the VBM metrics are straightforward adaptations of traditional DCF tools. As such they should yield the same predictions for project value (if properly used) as the traditional DCF analysis tools. The caveat "if properly used" is an important one, as we discuss here at some length.

In this chapter we examine the period-by-period performance of new capital investment projects using EVA, and we demonstrate that

GAAP depreciation can distort this performance metric. Switching to present-value depreciation (defined later) eliminates the source of the performance distortion but can give rise to a serious conflict of interest. Specifically, estimating present-value depreciation requires a firm's management to provide estimates of future cash flows, but these same employees are the ones whose performance is being evaluated.

Example Capital Investment Project

To illustrate the relationship between traditional measures of project value such as net present value (NPV), accounting return on invested capital (ROIC), internal rate of return (IRR), and the tools of value-based management, we consider the investment opportunity shown in table 7.1. The investment involves spending a total of $16,000 for plant and equipment plus an additional $2,000 for working capital. The plant and equipment are depreciated on a straight-line basis over seven years toward a zero salvage value. The $2,000 invested in working capital will be returned at the end of the project's seven-year life. The venture is expected to produce annual net operating profits after tax (NOPAT) of $1,200.77. Adding GAAP straight-line depreciation to NOPAT provides an estimate of the project's free cash flows of $3,486.49 per year. Finally, the opportunity cost of capital for the investment is 10%.

Traditional Measures of Project Value

Table 7.1 indicates that the project is a break-even proposition with a zero NPV and an IRR equal to the opportunity cost of capital. However, the accounting return on invested capital (ROIC) varies from 6.67% for year one to 28.02% for year seven. The increasing rates of return over time evidenced in ROIC reflect the fact that project income remains constant while the depreciated book value used to measure invested capital is declining. Does the fact that ROIC is below the 10% opportunity cost of capital during the first three years of the project mean that the project is destroying value during this time? Not necessarily. As we learned in chapter 3, ROIC calculated using GAAP accounting income and asset book values based on GAAP depreciation is an unreliable indicator of value creation since it fails to consider cash flow or the time value of money. Perhaps the new

Table 7.1. Traditional Measures of Project Value: ROIC, IRR, and NPV

	0	1	2	3	4	5	6	7
Net operating profits after tax (NOPAT)		$1,200.78	$1,200.78	$1,200.78	$1,200.78	$1,200.78	$1,200.78	$1,200.78
Depreciation expense (straight-line method)		2,285.71	2,285.71	2,285.71	2,285.71	2,285.71	2,285.71	2,285.71
Cash flow from operations		3,486.49	3,486.49	3,486.49	3,486.49	3,486.49	3,486.49	3,486.49
Plant & equipment	(16,000.00)							(0.00)
Working capital	(2,000.00)							2,000.00
Invested capital	(18,000.00)							
Free cash flow	(18,000.00)	3,486.49	3,486.49	3,486.49	3,486.49	3,486.49	3,486.49	5,486.49
Book value of capital	18,000.00	15,714.29	13,428.57	11,142.86	8,857.14	6,571.43	4,285.71	2,000.00
ROIC		6.67%	7.64%	8.94%	10.78%	13.56%	18.27%	28.02%
IRR	10.00%							
NPV	($0.00)							

tools of VBM would do a better job of measuring period-by-period value creation. Let us see.

Using EVA to Evaluate Project Value Creation

Table 7.2 contains estimates of the annual EVA for the example project introduced in table 7.1. How are we to use EVA to determine whether the project is a worthwhile investment? Is value being created in each period of the project's life? The answer to the first question is straightforward. A project's NPV of free cash flow always equals the present value of its EVA. Since our example project's NPV based on free cash flow—and the present value of EVA—are both equal to zero, the project neither creates nor destroys shareholder value. However, the answer to the second question about the analysis of period-to-period performance is problematic. This point is obvious when we look at our example project's EVA over time. For instance, in year one, EVA begins at–$599.23 and then increases each year until it reaches $772.20 in year seven. Thus, the annual EVA sends mixed signals. In the first three years we would conclude that the project destroys value, but in the last four years we deduce that it contributes positively to firm value. Only by considering all of the EVAs over the life of the project can we decide whether it is worthwhile. The problem we are encountering in interpreting the year-to-year EVA is directly related to the changing ROIC that we noted earlier in table 7.1. Recall that the returns increase over time as a direct result of depreciating invested capital. The link between ROIC and EVA is transparent in the following formulation of EVA (introduced in chapter 5):

$$EVA_t = (ROIC_t - k_{wacc})\ Invested\ Capital_{t-1} \qquad (7.1)$$

Recall that invested capital is equal to the depreciated book value of the investment based on GAAP accounting depreciation. In our example, $ROIC_t$ (and correspondingly EVA) increases over time due to the decreasing book value of the investment. Consequently, even though the EVAs taken together provide an appropriate basis for evaluating the project's NPV, the period-by-period measures are distorted by the use of GAAP accounting depreciation.

Fixing the Problem

Al Ehrbar, a former partner at Stern Stewart and now head of EVA Advisers, appropriately traces the root of the problem with the use of an annual EVA to evaluate the period-by-period performance of

Table 7.2. Analyzing Period-by-Period Performance Using EVA

	0	1	2	3	4	5	6	7
NOPAT		$1,200.77	$1,200.77	$1,200.77	$1,200.77	1,200.77	$1,200.77	1,200.77
Depreciation		2,285.71	2,285.71	2,285.71	2,285.71	2,285.71	2,285.71	2,285.71
Plant & equipment	($16,000.00)							(0.00)
Working capital	(2,000.00)							2,000.00
Invested capital	($18,000.00)							
Firm free cash flow	($18,000.00)	$3,486.49	$3,486.49	3,486.49	$3,486.49	$3,486.49	$3,486.49	5,486.49
Book value of capital	$18,000.00	15,714.29	13,428.57	11,142.86	8,857.14	6,571.43	4,285.71	2,000.00
Capital cost		1,800.00	1,571.43	1,342.86	1,114.29	885.71	657.14	428.57
EVA		($599.23)	($370.65)	($142.08)	$86.49	$315.06	$543.63	$772.20
MVA = PV (EVAs)	($0.00)							
IRR	10.00%							
NPV	($0.00)							

new investments according to the GAAP depreciation calculation. He describes the problem as follows:

> For most companies, the straight-line depreciation of plant and equipment used in GAAP accounting works acceptably well. While straight-line depreciation doesn't attempt to match the actual economic depreciation of physical assets, the deviations from reality ordinarily are so inconsequential that they do not distort decisions. That's not true, however, for companies with significant amounts of long-lived equipment. In those cases, using straight-line depreciation in calculating EVA can create a powerful bias against investments in new equipment. That's because the EVA capital charge declines in step with the depreciated carrying value of the asset, so that old assets look much cheaper than new ones. This can make managers reluctant to replace "cheap" old equipment with "expensive" new gear. (Ehrbar 1998)

Stern Stewart recommends a heuristic approach to resolving this problem. Ehrbar described the specific procedure to us in a telephone conversation in 2001:

> If an investment's cash flows are slow at coming on line, Stern Stewart uses suspension accounting treatment or strategic treatment. The approach is similar to the concept of construction in progress used in the utilities industry. For instance, if we make an investment that will not come on line for three years, we would hold the capital investment in a suspense account for that time. However, we would accrue interest on the capital at the rate of the firm's weighted cost of capital while it is held in suspension, so that the capital is not free to management. Then the higher capital would be used in the EVA computations in later years when the cash flows are being realized.

Bierman (1988) and later O'Byrne (2000) have proposed an analytical solution to the problem of measuring period-by-period performance with EVA. They suggest the substitution of present-value depreciation for traditional GAAP depreciation. Specifically, present-value depreciation for the year is defined as the change in the present value of the project's future cash flows, where these are discounted using the internal rate of return. Thus, for year *t* we estimate as follows:

$$\textit{present value depreciation} = \sum_{T=t+2}^{7} \frac{FCF_T}{(1+IRR)^{T-1}} - \sum_{T=t+1}^{7} \frac{FCF_T}{(1+IRR)^{T}}. \tag{7.2}$$

Figure 7.1 illustrates this procedure for our example project. In year zero the present value of the project's free cash flow is $18,000, which declines to $16,314 by the end of year one. Thus, present-value depreciation for year one is the difference in these two present values or –$1,686. We apply Bierman's concept of present-value depreciation to our example problem and report the results in table 7.3. Here we find that the reconstituted EVAs are all equal to zero, which is consistent with the fact that the project yields a zero NPV. Furthermore, the present value of the revised EVA estimates is still equal to the project's NPV of zero.

Unequal Cash Flows and Positive NPV

Let us now consider an investment project that is similar to our earlier one, except that cash flow is no longer the same from year to year, and the NPV is positive. We have the same investment outlay of $18,000, of which $16,000 is for plant and equipment, and $2,000 is for working capital, which will be recouped at the end of the seven-year project. However, the project's free cash flow is now $3,365.71 for the first three years and increases to $4,168.51 for the remaining four years. The cost of capital remains at 10%. The present-value analysis for this project is shown in panel A of table 7.4, where we see that the project has an IRR of 12% and a positive NPV of $1,324.

In panel B of table 7.4 we calculate the project's EVA for each year using present-value depreciation as proposed by Bierman (1988) and demonstrated earlier in Figure 7.1. These EVAs are all positive, which is consistent with the fact that the project has a positive NPV. Also, the project's MVA equals its NPV, and all seven $ROIC_t$ estimates are equal to the project's 12% IRR.[1] Appendix 7A demonstrates the equivalence of MVA and NPV.

Summary

What we have learned about the use of VBM tools for new-project analysis is that single-period performance measures made popular by VBM vendors for the evaluation of the performance of ongoing firm operations can easily be misinterpreted and misused to evaluate the period-by-period performance of new investment opportunities. Nothing is wrong with single-period measures of performance such as EVA per se; the problem lies in the use of single-period measure as an indicator of value-creation potential for long-term projects. The simple answer is to

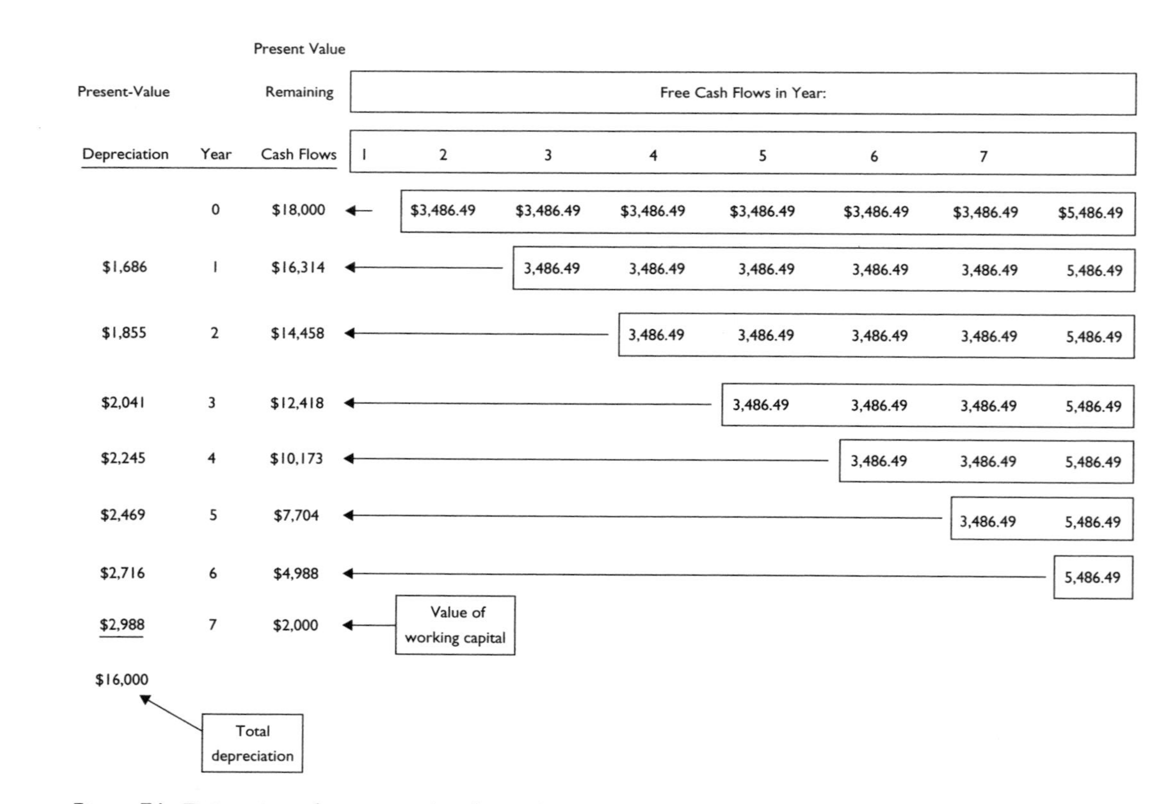

Figure 7.1. Estimation of present-value depreciation

Table 7.3. Revised Calculation of EVA Using Present Value Depreciation

	0	1	2	3	4	5	6	7
Book value of invested capital	$18,000.00	$16,313.51	$14,458.38	$12,417.72	$10,173.01	$7,703.82	$4,987.72	$2,000.00
Cash flow from operations		$3,486.49	$3,486.49	$3,486.49	$3,486.49	$3,486.49	$3,486.49	$3,486.49
Present value depreciation		1,686.49	1,855.14	2,040.65	2,244.72	2,469.19	2,716.11	2,987.72
NOPAT (revised)		$1,800.00	$1,631.35	$1,445.84	$1,241.77	$1,017.30	$770.38	$498.77
Capital cost (revised)		(1,800.00)	(1,631.35)	(1,445.84)	(1,241.77)	(1,017.30)	(770.38)	(498.77)
Return on invested capital (revised)		10.0%	10.0%	10.0%	10.0%	10.0%	10.0%	10.0%
EVA (revised using present value depreciation)		$0.00	$0.00	$0.00	$0.00	$0.00	$0.00	$0.00
IRR	10.00%							
MVA = PV (revised EVAs)	$0.00							
NPV	$0.00							

Table 7.4. Example with Unequal Cash Flows

Panel A. Project NPV and IRR								
Years	0	1	2	3	4	5	6	7
Free cash flows	$(18,000.00)	$3,365.71	$3,365.71	$3,365.71	$4,168.51	$4,168.51	$4,168.51	$6,168.51
Cost of capital	10%							
NPV	$1,323.94							
IRR	12%							

Panel B. Revised EVA Using Present-Value Depreciation								
Years	0	1	2	3	4	5	6	7
Present value of invested capital	$18,000.00	$16,794.46	$15,444.24	$13,931.98	$11,435.43	$8,639.28	$5,507.56	$2,000.00
Free cash flow		$3,365.71	$3,365.71	$3,365.71	$4,168.51	$4,168.51	$4,168.51	$4,168.51
Present-value depreciation		(1,205.54)	(1,350.22)	(1,512.26)	(2,496.55)	(2,796.15)	(3,131.72)	(3,507.56)
NOPAT		$2,160.17	$2,015.49	$1,853.45	$1,671.97	$1,372.36	$1,036.79	$660.96
Cost of capital		(1,800.00)	(1,679.45)	(1,544.42)	(1,393.20)	(1,143.54)	(863.93)	(550.76)
EVA		$360.17	$336.05	$309.03	$278.77	$228.82	$172.87	$110.20
Return on invested capital (NOPAT/beginning capital)		12%	12%	12%	12%	12%	12%	12%
MVA = PV (Revised EVAs)	$1,323.94							

use project NPV. Furthermore, NPV is completely consistent with EVA, where we consider the present value of all future project EVAs.

A modification of EVA does correct for the problems that arise in the use of traditionally defined EVA in project analysis. However, the fix comes at a high cost in terms of the required information inputs. The problem with using economic depreciation, as Bierman (1988) and O'Byrne (2000) recommend, is that the estimates required in implementing the system come from the firm's management. Since managerial compensation depends upon these estimates, managers would have an economic incentive to manipulate their estimates opportunistically. We discuss this problem more fully in chapter 8 where we review compensation issues in the context of VBM.

Appendix 7A

The Equivalence of MVA and NPV

Market value added (MVA) is the term Stern Stewart uses to measure the incremental value that a firm's management has added to the capital that has been invested in the firm. The calculation involves first computing the market value of a firm's equity and then adding the value of its liabilities. From this sum is subtracted an estimate of the total invested capital (book value of the firm's assets adjusted for a number of accounting practices that typically make this book value an underestimate of invested capital, e.g., the expensing of R&D costs). However, another interpretation of MVA assumes that the firm's market value is equal to the discounted present value of its future cash flows. We demonstrate that MVA is analogous to the net present value of the firm as a whole.

To illustrate the relationship between MVA and NPV, let us consider a single-period investment project in which a firm invests I_0 dollars in return for operating income at the end of the period equal to NOI_1. The firm pays tax at a rate T, the asset is fully depreciated in one period, leaving no residual or salvage value (i.e., depreciation expense for the period equals I_0), and the firm faces an opportunity cost of capital of k_{wacc}. In this simple setting, the year one EVA_1 can be defined as follows:

$$EVA_1 = NOI_1(1 - T) - k_{wacc}I_0. \tag{A.1}$$

Then,

$$MVA = \frac{NOI_1(1-T) - k_{wacc}I_0}{(1+k_{wacc})^1}. \tag{A.2}$$

To see the equivalence of MVA and NPV, note that the project's free cash flow (FCF) is defined as follows:

$$FCF_1 = NOI_1(1 - T) + I_0, \tag{A.3}$$

where depreciation expense equals I_0. Solving (A.3) for $NOI_1(1 - T)$ and substituting the result into (A.2) gives the following:

$$\begin{aligned} MVA &= \frac{FCF_1 - I_0 - k_{wacc}I_0}{(1+k_{wacc})^1} = \frac{FCF_1 - I_0(1+k_{wacc})}{(1+k_{wacc})^1} \\ &= \frac{FCF_1}{(1+k_{wacc})^1} - I_0 = NPV. \end{aligned}$$

The equivalence of MVA and NPV can easily be extended to multiple future periods.

Chapter 8

Incentive Compensation: What You Measure and Reward Is What Gets Done

> Academics and practitioners from a wide range of backgrounds agree that bringing about sustainable, productive changes in organizations is difficult. They disagree, however, on why this is the case. Consequently, they disagree on the most effective approaches to analyzing and solving organizational problems and on the most effective approaches to implementing solutions. At the heart of the disagreement are differences over the factors that motivate individuals to change their behavior. Behavioral changes on the part of individuals are required for organizational change, and compensation systems affect behavior. Thus, it is critical to consider the role that compensation systems play in the process of organizational change.
>
> —Karen Hopper Wruck, "Compensation, Incentives, and Organizational Change: Ideas and Evidence from Theory and Practice"

> Finally, we endeavor to ensure that our compensation program is perceived as fundamentally fair to all stakeholders.
>
> —Whole Foods Market 2007 Annual Proxy Statement

Every business faces the problem of motivating its employees to direct their efforts in ways that create value for the firm's owners. The problem is particularly acute for the non-owner-managed corporation since the owners (i.e., the stockholders) do not exercise direct control over the firm's operations. Two fundamental approaches can be taken. The first involves careful monitoring of employee behavior. Although this method can be effective, it becomes very cumbersome and consequently quite expensive to implement in large organizations, where decentralized decision making is required to make timely choices in the face of competition. Thus, the effectiveness of "watching over the employee's shoulder" is limited. The second approach involves developing a compensation policy that attracts, retains, and motivates high-performing personnel. This method forms the basis for value(s)-based management systems and is the subject matter of this chapter.

Before we proceed, we wish to point out that a compensation system is not limited to monetary payments. Wruck (2000) notes that human beings value and dislike more than monetary considerations within a business organization and that monetary consideration cannot capture everything that affects their motivation and incentives. However, for managers who are considering the redesign of their firm's compensation system, monetary rewards are particularly important and provide a productive place to start. The importance of money is not that individuals value it more highly than other types of rewards but that money is fungible; that is, it can be converted into whatever the employee values most of all. Thus, with the important caveat that compensation systems must deal with more than the allocation of monetary rewards, we focus our discussion on the financial component of a firm's compensation system.[1]

Young and O'Byrne (2001, 114–15) identify four basic objectives for a firm's compensation policy:

1. *Alignment:* To give management an incentive to choose strategies and investments that maximize shareholder value.
2. *Leverage:* To give management sufficient incentive compensation to motivate them to work long hours, take risks, and make unpleasant decisions, such as closing a plant or laying off staff, to maximize shareholder value.
3. *Retention:* To give managers sufficient total compensation to retain them, particularly during periods of poor performance due to market and industry factors.
4. *Shareholder cost:* To limit the cost of management compensation to levels that will maximize the wealth of current stockholders.

Even though each firm may describe its compensation policy objectives in a slightly different manner, these four objectives are universal.[2]

Moreover, they are important to the design of an effective compensation policy; however, value-based management systems have generally focused on the alignment and leverage objectives. Consequently, these form the primary basis for our discussion.[3] We do not, however, wish to turn a blind eye to the continual controversy that surrounds the topic of executive compensation cost. One cannot read a *Wall Street Journal* or a business magazine without seeing some references to outsized executive pay packages. We therefore include a discussion of compensation efficiency issues, in particular the fact that some forms of compensation are much less efficient and therefore much costlier than others. We also examine the way in which one firm, Whole Foods Market, is able to integrate a "fairness" component within its value(s)-based, management-oriented compensation. The political costs of ignoring fairness perceptions will likely become increasingly difficult for firms to bear.

Compensation plans vary in their complexity across firms, industries, and level of employee in a firm's hierarchy. Consider the typical CEO's compensation package, which Murphy (1999) describes as base salary, an annual bonus tied to accounting performance, stock options, and long-term incentive plans (including restricted stock plans and multiyear accounting performance plans). In very broad terms we can think of the base salary as fixed compensation in that it does not vary with a firm's performance. The remaining components are variable or at risk because they depend on certain measures of the company's performance. Value-based management systems are concerned with the design of measurement and pay-for-performance systems that determine the variable component.[4]

Most of this chapter is devoted to the use of the annual performance measures to determine the yearly bonus component of managerial compensation. It relates directly to the emphasis placed on it within the value-based management literature. One of the principal points we make is that a properly structured incentive compensation plan must create a culture of ownership in which the employees behave as if they are owners. We stress that, contrary to common belief, common stock and stock options are not the best way to develop this culture. Rather, bonuses that are determined by value-based metrics can fulfill this function while avoiding many of the problems associated with stock-based compensation. Second, we point out that annual pay-for-performance bonuses that are determined by the tools of value-based management can fail to provide a long-term value perspective. In fact, we demonstrate that these systems can lead to myopic investment decisions if they are not carefully designed. In addition, we discuss some of the mechanisms that have been introduced to mitigate this shortsighted behavior.

In the next section we discuss common errors that occur all too often in compensation plans at many firms. As we have just mentioned, compensation plans need to create an ownership culture in which the employee thinks and acts like an owner. In the following section we discuss how to do this and explain why stock options do not accomplish this goal. We then identify in more detail the three fundamental issues that every compensation program must address: the level, form, and composition of compensation. Within this section we discuss how to structure incentive plans that work well within a value(s)-based management framework. Finally we look at the compensation plan implemented at Whole Foods Market to illustrate that utilizing value-based metrics can be successfully integrated with CSR considerations.

All-Too-Common Mistakes

It would be quite comforting to believe that employees will work hard and do the right things without any kind of special rewards. A wealth of academic research, however, indicates that this utopian scenario simply does not exist, at least not in the vast majority of situations. Instead, we must rely on properly structured compensation contracts to guide and motivate the correct behaviors. Because compensation can be such a strong motivator, it is crucial, in the words of Jack Welch, to "get this one right":

> You have to get this one right. One time, I was surprised to see a great fourth-quarter revenue line and no income to go with it. I asked, "What the hell happened here?" "Well, we had a fourth-quarter sales contest, and everyone did a great job!" "Where's the margin?" "We didn't ask for margin."
>
> That's the simplest example of a universal problem: What you measure is what you get—what you reward is what you get. Static measurements get stale. Market conditions change, new businesses develop, new competitors show up. I always pounded home the question "Are we measuring and rewarding the specific behavior we want?" By not aligning measurements and rewards, you often get what you're not looking for.
>
> Jack Welch, former CEO of General Electric (2001, 387)

Unfortunately, this is not an easy endeavor, and there are a lot more ways to get it wrong than to get it right, with potentially disastrous consequences.

Correctly designed incentive compensation plans are essential to a firm's overall value(s)-based management program. These plans help ensure that the employees properly invest in and use the firm's assets wisely in order to build value. Improperly designed programs can do just the opposite. Remember, you get what you measure and reward, so be careful that you measure and reward the correct things!

We start this chapter by mentioning the things that do not work.[5] This may seem to be a backward approach; however, it is important to be up front about these mistakes, for they occur far too often. It is quite instructive to first clear the slate before discussing what we have learned about the proper way to structure incentive compensation within a V_sBM program.

The Wrong Performance Metric

The first (and probably the most common) mistake is to base the bonus on the wrong metric. Common metrics that often motivate behavior (as opposed to wealth building) include sales growth, net income, earnings per share, cash flow, return on assets, and market share. While any of these metrics will likely motivate employees to work toward increasing that metric, this same behavior may not effectively increase wealth, as Jack Welch learned when GE used sales growth as a performance metric.

By paying workers for making decisions and taking actions that mimic those of an owner, properly designed compensation plans make staff members think and act like owners rather than employees. Owners will likely be far more concerned about the bottom-line profits and the investment needed to attain them than about the top-line revenues. They will also be interested in the amount of their own capital that is tied up in the enterprise. Incentives should be designed to make employees reason this way.

Lack of Proper Training

Employees must understand the linkage between their behavior, the chosen performance metric, and wealth creation. Without this requisite knowledge they will likely fail to understand how their actions will not only be rewarded but also create value.

This failure to associate employee actions with desired outcomes is even more problematic at lower levels of the organization, where "line of sight" between actions and outcomes becomes less direct. *Line of sight* refers to a manager's belief that employees' actions have the power to bring about the required outcomes. While the workers in the trenches are obviously critical to the firm's performance, it may be difficult for these employees to see the impact of their efforts on the firm's stock price.

Lack of Exposure

Closely related to insufficient training is inadequate promotion of the incentive plan. Too often the firm designs an incentive plan in a vacuum and does little to communicate its existence. Instead of openly guiding employees and giving them feedback on how they are doing, the incentive plan remains largely ignored because it is relegated to the background.

One firm we worked with began prominently displaying graphs that charted key metrics on large posters for every employee to see. Suddenly water-cooler talk began revolving around how each employee could affect the metrics and how this would lead to bigger bonuses.

Overdependence on Short-term Performance

Value is based on how well the firm performs in the long run, yet many incentive plans reward only short-term performance. This can lead to behavior that puts too little attention on long-term investment and too much on short-term gaming such as accounting manipulation.

Enron and WorldCom are vivid examples of what can happen when short-term incentives to present a rosy picture become the focus of attention. A tremendous body of academic research has shown that the market ultimately values a company's ability to continually earn a return greater than its cost of capital. It is these future returns that create wealth, not short-term accounting magic. When incentives are based on simply exceeding some short-term measure of performance (such as earnings per share), especially if incentive pay is significant, it is surprising that there have not been many more Enrons. You get what you measure and reward.

Overreliance on Stock Options

Once thought of as the perfect compensation, stock options now appear to have been overused and even abused. While it is true that stock options may align the interests of managers and shareholders, this alignment is far from direct. In addition, stock options, once considered as basically free, can represent a very costly form of compensation.

We have more to say about this shortly. At this time we would like to quote Warren Buffet, who in his 1992 letter to the shareholders of Berkshire Hathaway questioned the firm's overuse of options based on the rationalization that they are not an expense and are therefore not costly: "If options aren't compensation, what are they? If compensation isn't an expense, what is it? If an expense does not get charged to earnings, where in heaven's name should it go?"[6]

Limiting Incentive Pay to Top Managers

Incentive plans at many companies are limited to the top managers or certainly concentrated mostly at the top levels. Most employees receive either no incentive pay or at best only an insignificant portion of their compensation as incentive pay. In addition, the incentive pay that does exist (even at the top level) is often too small to provide the motivation to make the truly tough decisions.

As we demonstrate later in this chapter, if the bonus is structured as a percentage of performance improvement rather than of the level of performance, there is no reason to cap incentive pay at some low level.[7] In essence, the bonus funds itself since it comes from additional wealth that is being created by the efforts of those receiving the incentive pay. Therefore, the larger the bonus, the better the plan is for all.

Creating a Culture of Ownership

Consider how people typically treat a rental car compared to the way they care for their own vehicle. They simply act differently when they are the owner. Because of this, it is important for the company to create a culture of ownership, in which the employees act as owners of the company, not as merely employees. Properly designed compensation packages can help create such a culture, whereas paying just a fixed wage or salary keeps employees acting as such. Nothing can be more important for a value-based management program than to convince employees to act like owners.

The primary vehicle most companies use to foster this ownership culture is to arrange for employees to hold company stock or stock options. After all, what better way to have workers act and think like owners than to make them owners? Unfortunately, stock-based incentives do not come without their own set of problems.

The first and most significant one from an incentive viewpoint is that stock prices provide only an aggregate value for the entire company. Rarely, however, will this represent a proper measure for valuing the contribution of an employee who is usually responsible for only a small part of the organization. Many, if not most, companies are composed of different divisions, product lines, customer segments, and so on. As Bennett Stewart III (2002, 3) writes, "Stock and stock options are like issuing only one report card for the whole class of students. The best students (or employees) are demoralized, and the worst ones are thrilled to coast on everyone else's efforts."

Economists refer to this as a *free-rider problem.* These employees had a free ride since they did nothing but ride on the coattails of others whose actions drove the share price upward. Since each individual likely has little overall effect on the company's stock price, there is scant motivation to work extra hard. Very few of the rewards accrue to you, and your efforts mostly serve to benefit all of the others, who may not be working nearly so hard.

In order for incentives to work as intended, it is essential to break the company down into smaller units. This way each employee's incentive compensation can be tied to an area in which that person has a stronger line-of-sight responsibility. Only then can a culture of ownership be established.

An additional problem with stock options is that, contrary to popular belief, they are a very costly and inefficient form of compensation. The popular belief is that since no cash is initially paid out (and potentially some will actually come in when the options are exercised) and since (until recently) accounting rules did not require options to be expensed on the income statement, stock options represented little or no cost to the company. Of course, this is complete fantasy.

First let us consider the arguments that the firm does not give up cash and that the options represent a potential source of ready money to the firm. The company will be required to surrender to the option holders a share of company stock if they choose to exercise their option. This will occur only if the value of a share of stock is greater than the cash exercise price. Therefore, the company is giving up something of greater value than the cash it is receiving. It must either go out and purchases shares in the open market for this distribution (in which case it is expending actual cash), or it must issue new shares, which dilute the value of the existing shares. Either way, a real cost is being borne by the company and hence by the existing shareholders.

Equally fallacious is the argument that no cost is involved if the options do not have to be expensed on the income statement. Accounting rules should not be relied upon to determine economic reality. They are the result of a political process that in many circumstances defies logic and is often dominated by lobbying efforts. We doubt many would feel that rent payments would not be an economic cost if, for some reason, accounting rules were changed to exclude this expense from the income statement.

Stock options are not only costly but also a very inefficient form of compensation. They are far different from cash when it comes to value perceptions. Because it has a uniform applied value, cash is easily valued. One dollar of cash given up by the firm is worth—and valued as—one dollar by the employee. Not so with stock options, however. Numerous studies show that an employee will value stock options at a value that is far less than the cost of those options to the issuing company. Hall and Murphy (2000, 2002) show that, with reasonable assumptions about risk aversion and diversification, employees value options that have an exercise price equal to the existing market price of the stock at only half their cost to the issuing firm.

A final problem with equity-based compensation is that, while stock prices may provide the company's ultimate score, they certainly do not give the employees the information they need to know in order to score. So many things, most of which are outside the control of any individual (or the entire company for that matter), affect the price of the company's stock. Employees should be judged by concrete metrics that they can influence and that can be shown to affect stock price—these are the *value drivers* we refer to in chapter 4).

Recent academic evidence supports the argument that stock options are quite problematic as a form of incentive compensation. Gerald Sanders and Donald Hambrick (2007) followed company results over a seven-year period and noted the relationship between the company's investment and financial performance and the percentage of CEO compensation in the form of stock options. They found a negative (inverse) correlation between stock options and firm performance; that is, company performance improved as the use of stock options decreased. Sanders and Hambrick observed that "high levels of stock options appear to motivate CEOs to take big risks, or to swing for the fences," but noted that "they strike out more often than they hit home runs" (1073). The problem, they maintain, is that the options do not motivate them to consider the downside.

Although it is very unlikely that stock or stock options can accomplish this line-of-sight responsibility, this does not mean they should be abandoned. Rather, they can serve as an excellent complement to a well-designed bonus plan. In particular, stock-based compensation can help lengthen the employee's horizon, something we discuss later in this chapter.

The primary incentive should be a bonus tied to individual and business-unit performance. Such a properly designed bonus can foster a culture of ownership since it focuses on line-of-sight accountability without suffering so acutely from the many problems associated with stock-based incentives. In addition, tying a portion of the bonus to both companywide and business-unit performance fosters an environment of teamwork. In order for bonus plans to overcome the problems associated with stock incentives and at the same time create an ownership culture, they must be strategically developed to generate an ownership interest in the company. They must be linked not only to the organization's overall well-being but also to long-range wealth creation rather than merely short-term performance.

We next turn our attention to items that a firm's compensation policy must address.

Determining a Firm's Compensation Policy

A firm's compensation policy must deal with three fundamental issues: (1) what level of compensation to pay; (2) how to connect the compensation to performance (i.e., the functional form); and (3) how to pay the compensation (i.e., its composition), including consideration of cash payments versus benefits (e.g., insurance, working conditions, leisure time) and cash versus equity (stock options and grants) for incentive pay.[8]

What Should the Level of Compensation Be?

At a minimum, a competitive labor market requires that executives be paid what they might earn in a comparable job elsewhere. In essence, the level of compensation a firm pays will determine the quality and quantity of workers the firm can attract. Furthermore, market forces that are outside the firm's control largely determine the level of compensation it can pay. It is standard practice for firms to use compensation survey data to determine the level of compensation they pay to their employees. For example, Dana Corporation's 2005 proxy statement says, "In making our compensation decisions, we consider competitive market data.... This data [*sic*] compares Dana's compensation levels, and the relationship between Dana's compensation levels and corporate performance, with those of companies in a peer group.... Currently, there are 17 companies in the peer group" (14). According to Murphy (1999), compensation surveys are used nearly universally as a means of determining base salaries.

The proponents of value-based management argue that paying a competitive level of compensation is not sufficient to ensure the creation of shareholder value. Specifically, they argue, one must link pay to performance. The fundamental premise underlying value-based management systems is very simple: What a firm measures and rewards gets done. So, if your goal is to create firm value, one should select a performance metric that is consistent with that objective and use it to determine compensation.[9] Thus, value-based management systems focus on the second component of compensation policy, which is the functional form of the compensation.

How Should Pay Be Linked to Performance?

Functional form refers to the relationship between pay and performance. The sensitivity of pay to performance depends upon two attributes of the firm's compensation program: (1) the fraction of total compensation that is tied to performance and (2) the formula used to relate pay to performance. There are no hard and fast rules for determining the mix of variable and fixed compensation. However, in practice we observe that a firm's highest-ranking employees have a larger fraction of their total compensation that is at risk and dependent on the company's performance. For most firms this simply mirrors the responsibilities of their top managers and their ability to control the company's performance. Furthermore, because the at-risk or variable component of compensation is the key to determining the sensitivity of compensation to performance, we focus on this factor as an influence on incentive compensation.

Formula for Determining Incentive Pay

The procedure to determine the level of incentive compensation is similar regardless of the particular performance measure that is chosen.[10] Let us begin by looking at an unbounded incentive compensation payout formula:

$$\text{Incentive Pay} = \begin{pmatrix} \text{Base} \\ \text{Pay} \end{pmatrix} \begin{pmatrix} \text{Fraction} \\ \text{of Pay} \\ \text{at Risk} \end{pmatrix} \left(\frac{\text{Actual Performance}}{\text{Target Performance}} \right). \tag{8.1}$$

Since there are no limits specified in this case as to the maximum or minimum levels, the incentive pay is unbounded. Also, we see that the incentive compensation in equation 8.1 is a function of the amount of the employee's compensation that is at risk or subject to the firm's performance (the product of base pay and the fraction of pay at risk) and its actual performance for the period relative to a target level of performance. In practice, this basic system might be the same for all employees but differ in terms of the level of employee base pay and the percentage of that base pay that is at risk or subject to incentive compensation.

How does the basic model of incentive pay work? Consider the case of an employee whose base pay is $50,000 plus an additional 20% of this base pay (or $10,000) is at risk (i.e., dependent on the firm's performance). We define base pay as the employee's fixed compensation or

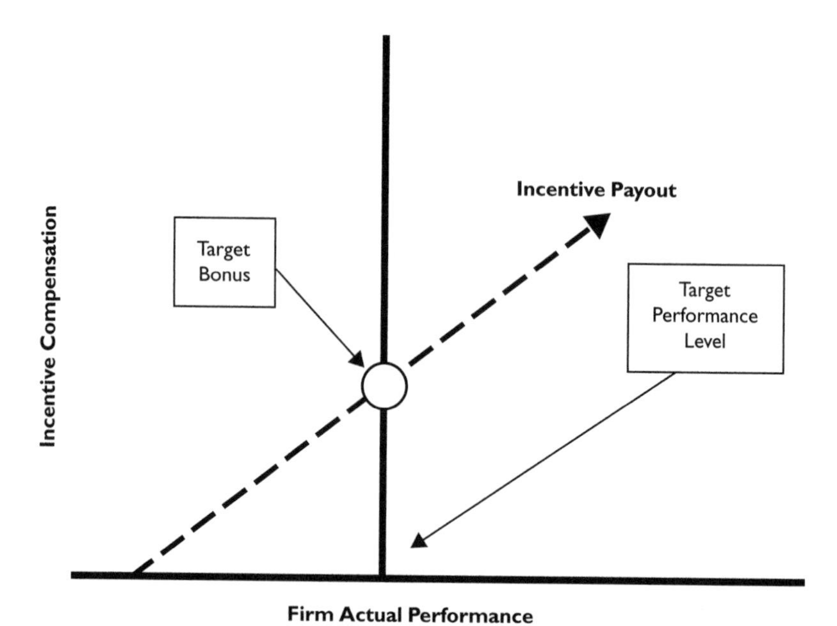

Figure 8.1. Pure (unbounded) incentive pay-for-performance system

salary, which does not vary regardless of the firm's performance. Thus, the employee's total compensation will equal the base component of $50,000 plus some fraction of the at-risk pay. If we assume that the ratio of actual to target performance is 1.1, then the employee's incentive or at-risk pay for the year is $11,000, and total compensation for the period equals $61,000.[11] Alternatively, if the firm's performance is just equal to the target level of performance, then incentive compensation is $10,000, and total compensation is $60,000.

Figure 8.1 illustrates how incentive compensation varies with a firm's performance. In this example incentive compensation is *unbounded* since it varies directly with actual performance relative to target performance with neither a floor (minimum) nor a cap (maximum).[12] Such a system provides the firm's employees with an incentive to improve their performance regardless of the company's own in such a system.

Most businesses, however, do not use an unbounded incentive pay program. Instead, they employ a system that provides for a minimum or threshold level of performance (in relation to the target level) before the incentive plan kicks in and a maximum level of performance (again in relation to the target), above which no incentive pay is rewarded. These minimum and maximum performance levels are sometimes referred to as "golfing points" because of the adverse incentives that they have on employees' work effort.

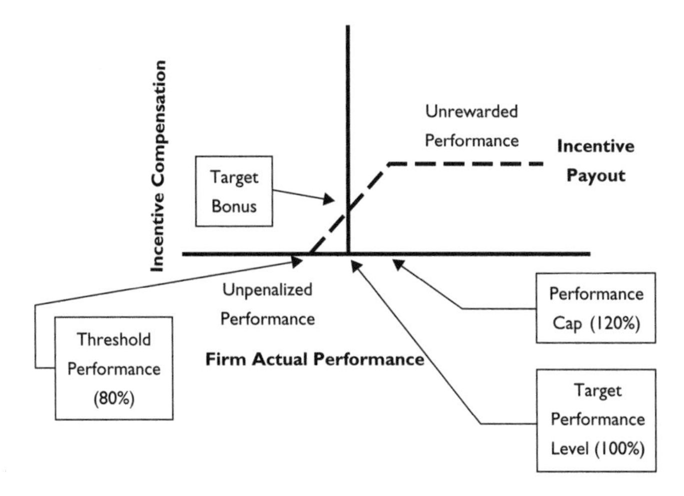

Figure 8.2. An 80/120 (bounded) incentive pay-for-performance system

Figure 8.2 contains an example of an 80/120 plan, for which the minimum (or threshold) level of performance for which incentive compensation is paid equals 80% of the target level of performance. The maximum performance for which incentive pay is rewarded is 120% of the target performance. Consequently, incentive compensation is paid only for performance levels that fall within the 80–120 range. In addition, for a wide range of firm performance, no incentive pay is awarded (i.e., performance greater than 120% or less than 80% of the target level). Thus, this type of program has incentive effects that are limited to the range of performance with which the payout varies.

Bounded incentive compensation plans, Zimmerman (2006) argues, can provide perverse incentives to managers. For example, as it becomes obvious that a firm's performance for the evaluation period (usually one quarter or one year) will fall below the 80% threshold, employees have no direct pay-for-performance incentive to work harder in order to enhance the firm's performance during the remainder of the evaluation period. In fact, they have an incentive to reduce current-period performance even further in the hope that it will lower the target level for the coming period. In addition, reducing current-period performance may allow the shifting of some or all of the lost performance to the following evaluation period, when the employees hope to be rewarded for it.

A similar perversity arises on the upper end of the performance spectrum. For example, if it looks like the firm's performance is going to surpass the maximum payout level, then the employees once again feel no incentive to improve their own performance. First, they are not paid for performance beyond the maximum, and second, they may be able to postpone business until the coming period, when they will be able to count it toward incentive pay.[13]

A growing body of evidence suggests that managers do behave opportunistically in reaction to bounded compensation plans. It suggests that managers bank their earnings when it appears that they will surpass the performance cap, but there is no evidence that they take an earnings bath when threshold performance is unlikely (Degeorge, Patel, and Zeckhauser 1998; Gaver, Gaver, and Austin 1995; Healy 1985; Holthausen, Larcker, and Sloan 1995).

Single-Period Performance Measures and Managerial Incentives

To this point we have referred to a firm's performance without specifying how to measure it. A wide variety of performance metrics might be used, including accounting-based measures such as earnings, earnings growth, and revenue growth, in addition to the value-based management system metrics. However, virtually all of these are based on the results of a single period's historical performance. This fact raises a very serious problem that Jensen (1998, 354–55) has summarized in the following way:

> Because EVA is a flow measure, it does not solve the capital value problem. This means that if future annual EVA of a project is sufficiently large, it will pay a company to take a project whose early years' EVA is negative. Market value, the discounted value of net cash flows less the investment required to generate them, is the appropriate value to maximize. Thus, while EVA is the best flow measure of performance currently known, it is not the universal answer to the search for the perfect performance measure.

This problem can create adverse incentives for managers whose decisions are based on personal financial inducements over a decision horizon that is shorter than the life of the projects under consideration.

Managerial Decision Horizon and the Use of EVA

We can illustrate the nature of the "stock versus flow" problem noted in the preceding quotation by using the example investment project found in table 8.1 (similar to those discussed in chapter 7). We assume an initial investment of $18,000, consisting of an outlay of $16,000 for plant and equipment, which will be depreciated on a straight-line basis ($2,285.71 per year) over a seven-year life, and $2,000 of working capital, which will be recouped at the end of the project. The project has a 10% required rate of return.

We begin our analysis of the project's cash flows and its value in Table 8.1 with its expected net operating profits after taxes (NOPAT). Observe that NOPAT is the same throughout the first three years of the project's life, then increases to a higher level for the final four years. We then add back depreciation to compute free cash flow. Note that book capital declines over time as the plant and equipment are depreciated, and the capital cost each year is the beginning book capital (ending capital in the prior year) times the cost of capital (10%). The results are as follows:

- The project's net present value is $132.81.
- Return on invested capital is equal to NOPAT divided by the beginning capital and grows from 6.33% to 30.80% over the life of the project.
- Economic value added is NOPAT less the capital cost. It is initially negative but increases over the life of the project from ($660.00) to $891.43.
- EVA based compensation is 1% of EVA for the period.
- Cumulative present value of the compensation is the sum of the present values of the annual EVA compensation for one year, two years, and so forth. If a manager's horizon were four years, then the cumulative present value impact of accepting the project on his bonuses is ($9.68), whereas would rise to $1.33 for a seven-year horizon.

Since the project has a positive net present value of $132.81, it is expected to create shareholder value. However, the EVAs for the project are negative for the first three years. Consequently, if incentive compensation is based on EVA, the project has a negative impact on managerial compensation. This fact is depicted graphically in figure 8.3, where we plot the sum of the present values of the annual bonuses for each year.[14]

Consider the financial incentive of a manager who is considering such a project but does not plan to remain with the firm for more than, say, three or four years.[15] This individual will not want to undertake the project even though it offers a positive net present value if EVA is used to evaluate annual performance. This is due to the fact that the project's annual performance metrics in any given year do not reflect the value of

Table 8.1. Example: EVA-Based Incentive Compensation for a Positive NPV Project

	0	1	2	3	4	5	6	7
Net operating profit after taxes		$1,140.00	$1,140.00	$1,140.00	$1,320.00	$1,320.00	$1,320.00	$1,320.00
Depreciation		2,285.71	2,285.71	2,285.71	2,285.71	2,285.71	2,285.71	2,285.71
Plant & equipment	$(16,000.00)							(0.00)
Working capital	(2,000.00)							2,000.00
Invested capital	(18,000.00)							
Firm free cash flow	(18,000.00)	3,425.71	3,425.71	3,425.71	3,605.71	3,605.71	3,605.71	5,605.71
NPV	$132.81							
Book capital	18,000.00	15,714.29	13,428.57	11,142.86	8,857.14	6,571.43	4,285.71	2,000.00
Capital cost		1,800.00	1,571.43	1,342.86	1,114.29	885.71	657.14	428.57
Return on invested capital		6.33%	7.25%	8.49%	11.85%	14.90%	20.09%	30.80%
EVA		(660.00)	(431.43)	(202.86)	205.71	434.29	662.86	891.43
EVA-based compensation (1% of EVA)		(6.60)	(4.31)	(2.03)	2.06	4.34	6.63	8.91
Cumulative PV of EVA compensation		($6.00)	($9.57)	($11.09)	($9.68)	($6.99)	($3.25)	$1.33

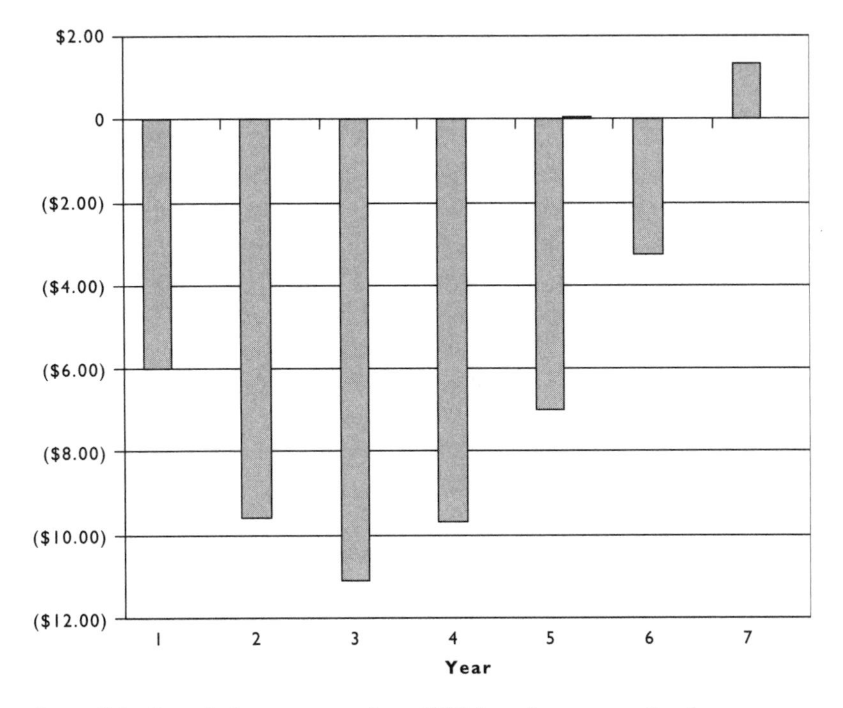

Figure 8.3. Cumulative present value of EVA-based compensation for a positive NPV project

future performance. As a result, a manager who is compensated based on EVA and with a decision horizon of less than six years will reject the project even though it offers a positive NPV.

It is important to note, however, that when the decision horizon is equal to the life of the project under consideration, the manager will be motivated to make decisions that are consistent with NPV. That is, the cumulative present value of the manager's bonuses based on EVA are positive for an administrator with a seven-year horizon, indicating that this individual will want to undertake the positive NPV project. This result is not surprising since, as we learned in chapter 7, the present value of project EVAs over the entire life of the project is equal to its NPV. Thus, the cumulative present value of future bonuses over the entire life of the project is simply a percentage (1% in our example) of NPV. Correspondingly, if NPV is positive, the cumulative present value of the manager's bonuses over the life of the project will also be positive, and conversely.

It can similarly be shown that equally perverse incentives can exist with a negative NPV project that promises high returns in the early years and lower returns in the later ones. In such a situation just the opposite result from the preceding example is quite likely to occur. Namely, because EVAs will be positive in the early years, the negative NPV project will be undertaken.

To summarize, when a single-period metric such as EVA is used to measure performance, managers with horizons shorter than the life of the project can have a financial incentive to behave counterproductively. Specifically, they may find that they will have an incentive to accept some projects that offer good near-term prospects but have a negative NPV and to reject positive NPV projects that have good long-term prospects but provide little cash in their early years. The primary source of the problem, of course, is the length of the manager's decision horizon.

Extending Managerial Horizons

One method that has been suggested for addressing the problem of managerial decision horizons and the single-period nature of the EVA measurement is to use a bonus bank system.[16] A bonus bank entails paying out employee bonuses over an extended period of time. For example, with a three-year bonus bank, the bonus earned this year would be paid out one third this year, one third next year, and one third the following year.[17] A bonus-bank system pushes the manager's decision horizon out to the length of the bonus-bank period since poor performance in the future will reduce the employee's bonus-bank balance. However, to be completely effective the bonus bank must extend the manager's decision horizon over the entire life of the projects being analyzed. In practice, bonus banks seldom extend beyond five years.

The bonus bank can provide additional help in mitigating an underinvestment tendency if the performance metric is based on improvements rather than levels. Cost cutting and failure to invest in the future can certainly lead to higher current performance. This type of behavior, however, will not be conducive to performance improvement. Therefore, even though a large bonus may be earned in the current year, poor performance in the future will prevent the current bonus from being realized.

Another solution that has been proposed to mitigate underinvestment associated with EVA is to set up suspense accounts to hold the investment off the balance sheet. This serves much the same function as present-value depreciation (chapter 7). Utilizing this approach, the investment is gradually brought back onto the balance sheet (and therefore into the capital charge calculation) over time instead of being charged at the full rate from the beginning.

A third approach, involving the use of separate value drivers in addition to the primary value-based metric, can also be used to extend the managerial horizon. Essentially, under this approach the bonus is potentially based on multiple metrics (some of which may be nonfinancial in nature) that often represent leading indicators of future value creation. For example, if new product development is an important contributor to the firm's value creation, a metric associated with it could be included in the bonus calculation.

Finally, stock-based compensation, discussed earlier in this chapter, can extend the managerial horizon because it provides incentives for long-term value creation. As we explained earlier, stock options suffer from a lack of line of sight; however, they should be included as part of an overall compensation package because they provide payoffs for investments that may depress short-term performance.

How Should Employee Compensation Be Structured?

In order to create a culture of ownership and convince employees to act like owners, you need to pay them like owners. This can be done with a properly designed bonus plan that has certain characteristics. The first and most important one is the proper alignment of employee and shareholder interests, which is best achieved when the employees' rewards are strongly linked to the creation of shareholder wealth. A proper value-based metric such as EVA is best suited for this purpose since shareholder wealth ultimately results from a firm's earning more than its cost of capital.

Simply choosing the right metric is not sufficient, however. It is also important to use that metric properly. This leads to the second characteristic of a properly designed bonus plan: basing the bonus on improvement instead of level, as we discussed earlier in this chapter. One should not be penalized by a low level of EVA—or some other performance measure—if one takes a poorly performing business unit to a higher level. Likewise, employees should not be rewarded just because they happen to inherit a high-performing unit where they simply coast along.

The third characteristic, materiality, combines the attributes of the first two features. It is not enough to have the right metric and use it properly if the amount of the potential bonus is insufficient to motivate the employee. It clearly needs to be material to the employee's well-being.

Closely associated with materiality is the characteristic of unlimited upside potential. Rather than capping an individual's bonus at some percentage (typically 120% of the targeted bonus), an unlimited upside more closely resembles ownership attributes. If the bonus is based on EVA improvement, it essentially becomes self-funding. Since the employees are sharing in any EVA enhancement, the larger the bonuses, the better off are both the employees and the company.

The website of Herman Miller, a leading global provider of office furniture, discusses how the company combines employee stock ownership and EVA to create a culture of ownership:

> Employee owners carefully monitor and know how their roles contribute to our profitability and daily reliability score—a score measuring our performance against customer expectations. Each month every employee reviews the numbers, particularly our EVA performance, a measure of our contributions to the long-term value of the company. We are owners, we think like owners, and we share in the fortunes of the business, like owners. We also work hard to understand our opportunities for long-term profitability and growth.

Fairness as an Additional Characteristic of a Firm's Compensation Policy

Regardless of how one feels about the economics of compensation policy, one simply cannot ignore the perceptions that exist. Simply put, executive compensation makes headlines, and, for the most part, these are not ones that firms like to see. A successful V_sBM program cannot ignore the political ramifications of the public's perceptions. As we demonstrate in chapter 6, corporate social responsibility makes good business sense. The major benefits of a strong CSR image are customer support and reduced governmental scrutiny.

Whole Foods Market was introduced earlier in chapters 1 and 6, where we indicated that the firm successfully practices value(s)-based management through the integration of an ambitious CSR program within a disciplined VBM framework. Whole Foods believes fairness should be an important part of its compensation. The appendix to this chapter presents the compensation discussion and analysis from the Whole Foods' 2007 proxy statement. First note that the objectives of the Whole Foods' compensation program are consistent with those illustrated in this chapter. Specifically, Whole Foods mentions the following goals:

1. to attract and retain qualified, energetic team members who are enthusiastic about the company's mission and culture
2. to provide incentives and reward all of the team members for their contribution to the company
3. to promote an ownership mentality among key leadership and the board of directors
4. to ensure that our compensation program is perceived as fundamentally fair to all stakeholders

The first three items should look quite familiar; however, the fourth item is an area of distinction within Whole Foods' compensation plan. The proxy statement provides further detail on this point. Whole Foods limits the maximum salary for any executive to at most a multiple of nineteen times the average annual wage within the company. In 2006 this amounted to a cap of $607,800. Whole Foods initiated this limit as part of its commitment to stakeholder equity.

At first this type of compensation, with its emphasis on the principle of stakeholder equity, may seem at odds with value-based management. Nonetheless, Whole Foods certainly does not believe this to be the case. All of the executive officers of Whole Foods participate in an incentive compensation plan based primarily on EVA improvement. According to the proxy statement, "The EVA bonus is included in compensation to align the financial incentives with the interests of our shareholders, which we believe is primarily the growth and return on invested capital."

Summary

A firm's compensation policy constitutes a critical component of its internal control system. That is, where owners (stockholders) do not manage the firms they own, the firm must have in place an internal control system that monitors employee performance and rewards those efforts that increase the firm's value. In this chapter we have discussed a fundamental component of every firm's internal control system, its compensation program. The basic paradigm espoused by the proponents of value-based management is that what a firm measures and rewards will get done. Consequently, if the firm is to be run so as to enhance its value, the compensation program must measure employees' activities that contribute to this goal and reward them. In essence, the compensation plan should pay employees to think and act like owners.

Employee compensation at most firms consists of both a fixed and a variable component. The latter is generally tied to some measure of the company's performance that is connected to its primary goal (maximizing the firm's value in this instance). Thus, incentive-based compensation (i.e., pay-for-performance or at-risk pay) is the compensation component of interest when designing a value(s)-based management system.

The level of incentive pay is usually linked to a comparison of actual and target (planned) performance. However, many firms limit the range of performance over which incentive pay is rewarded, but this can have an adverse effect on employees' motivation. Consider, for instance, the employees' response to a situation in which the firm's actual performance comes in below or above the threshold level for which incentive pay will be awarded. One solution to the perverse effects of floors and caps on incentive compensation is to remove them and create a long-term bonus bank into which employee incentive pay is placed and then paid out over a three-to-five-year period. The purpose of the bonus bank is to counter the employees' incentive by focusing on near-term performance at the expense of longer-term performance.

Once the level of incentive pay has been determined, its form and composition remain to be worked out. The key is to base incentive compensation on a metric that fosters a culture of ownership. While equity-based compensation is often promoted with this intent, we find that it is fraught with problems. The primary concern is a lack of line of sight between stock-based compensation and an individual's action (i.e., employees are unable to see how their actions can directly impact the stock price), which can lead to a free-rider problem. Another deficiency of stock-based compensation is its relatively costly and inefficient nature. Paying with equity can, however, lengthen the employee's time horizon. Therefore, we argue that stock-based compensation should be included in the compensation package; however, the annual bonus, when determined by a value-based metric such as EVA, should be the primary element.

An addition characteristic that firms must be aware of is the public's perception that executive compensation has become unrealistically excessive. A complete V_sBM program must consider these political implications because such perceptions can and do effect a firm's reputation and ultimately its wealth-creating ability. One mechanism for addressing them is to consider fairness criteria when designing a compensation program.

Appendix 8A

Whole Foods Market Executive Compensation Discussion and Analysis

Objectives of Compensation Program

The primary objective of our compensation program, including our executive compensation program, is to attract and retain qualified, energetic Team Members who are enthusiastic about the Company's mission and culture. A further objective of our compensation program is to provide incentives and reward each Team Member for their contribution to the Company. In addition, we strive to promote an ownership mentality among key leadership and the Board of Directors; our Corporate Governance Principles provide that it is the policy of the Board of Directors to encourage all directors to maintain an outright investment in the Company equal to the total estimated cash compensation received for serving on the Board of Directors over three years. Finally, we endeavor to ensure that our compensation program is perceived as fundamentally fair to all stakeholders.

What Our Compensation Program Is Designed to Reward

Our compensation program is designed to reward teamwork and each Team Member's contribution to the Company. In measuring the executive officers' contribution to the Company, the Compensation Committee considers numerous factors including the Company's growth and financial performance through reference to the following metrics. All of our executive officers participate in an incentive compensation plan based primarily on improvement in economic value added ("EVA"). EVA is the primary basis for the Company's financial decision-making tools and incentive compensation systems. In its simplest definition, EVA is equivalent to net operating profits after taxes minus a charge for the cost of capital necessary to generate those profits. High EVA correlates with high returns on invested capital. The incentive compensation paid to the executive officers for fiscal year 2006 was based upon the incremental improvement in the Company's overall EVA, the number of new stores opened or acquired during the fiscal year, and the number of new stores opened with total development costs within the development budget minus a charge for the new stores opened with costs in excess of the development budget during the fiscal year. Fiscal year 2006 incentive compensation averaged approximately 49% of the total cash compensation earned by the executive officers.

Regarding most compensation matters, including executive and director compensation and the Company's salary cap, our management provides recommendations to the Compensation Committee; however, the Compensation Committee does not delegate any of its functions to others in setting compensation. We do not currently engage any consultant related to executive and/or director compensation matters.

Stock price performance has not been a factor in determining annual compensation because the price of the Company's common stock is subject to a variety of factors outside our control. The Company does not have an exact formula for allocating between cash and non-cash compensation. Other than EVA pool and Benefit Hours pool balances (see below), compensation is generally paid as earned.

Elements of Company's Compensation Plan and Why We Chose Each (How It Relates to Objectives)

Annual executive officer compensation consists of a base salary component and the EVA incentive component discussed above. It is the Compensation Committee's intention to set total executive cash compensation sufficiently high to attract and retain a strong motivated leadership team but not so high that it creates a negative perception with our other stakeholders. The EVA bonus is included in compensation to align the financial incentives with the interests of our shareholders, which we believe is primarily the growth and return on invested capital.

Each of our executive officers receives stock option grants under the Company's stock option plan. All of our 40,000+ fulltime and part-time Team Members are eligible for stock option grants through Annual Leadership Grants, which recognize and incentivize Team Member performance, or through Service Hour Grants once they have accumulated 6,000 total service hours (approximately three years of employment). Approximately 94% of the stock options granted under the plan since its inception in 1992 have been granted to Team Members who are not executive officers or regional presidents. We believe that through our broad-based plan, the economic interests of our Team Members, including our executives, are more closely aligned to those of the shareholders. Other than Service Hour Grants, the number of stock options granted to each executive officer is made on a discretionary rather than formula basis by the Compensation Committee. Other than

Service Hour Grants, each executive officer receives the same number of stock option grants.

We also have a policy that limits the total cash compensation paid to any Team Member in each calendar year. The compensation cap is calculated each year as an established multiple of the average cash compensation of all full-time Team Members employed during such year. For fiscal year 2006, the Company increased the cap from 14 to 19 times the above described average. Employee benefits, stock options and any other form of non-cash compensation, such as the 401(k) match, are not counted in determining and applying the salary cap. Payouts under any EVA Incentive Compensation Plan ("EVA Plan") fall within the scope of the Company's salary cap policy. Per the EVA Plan, amounts are contributed annually to a "pool" for each Team Member based on EVA results. A portion of the annual EVA pool contribution may remain in the pool, and a portion is paid out annually. Annual payouts are calculated as 100% of the pool up to certain job-specific dollar amounts plus a portion of the excess. If the EVA bonus to be paid out will cause the Team Member's cash compensation to exceed the annual salary cap, the amount above the cap is forfeited; the full amount which would otherwise be paid out (including amounts not actually paid to the Team Member) is still subtracted from the pool balance. The accumulated balance in any Team Member's pool account is limited to the amount of the salary cap. Team Members may take time off without pay in order to reduce their salary earned and increase the amount of bonus that can be paid within the cap. The salary cap does not apply in the Team Member's year of termination or retirement.

The salary cap relates to the Company's commitment to stakeholder equity as a principle. The following is the salary cap calculation for the Company's past eight fiscal years:

Year	Average Hourly Wage	Average Annual Wage	Average Multiple	Salary Cap
1999	$12.36	$25,709	10	$257,000
2000	$12.84	$26,707	14	$373,900
2001	$13.46	$27,997	14	$391,900
2002	$13.69	$28,479	14	$398,700
2003	$14.07	$29,266	14	$409,700
2004	$14.66	$30,493	14	$426,900
2005	$15.00	$31,200	14	$436,800
2006	$15.38	$31,990	19	$607,800

How the Company Chose Amounts and/or Formulas for Each Element

Each executive's current and prior compensation is considered in setting future compensation. In addition, we review the compensation practices of other Companies. To some extent, our compensation plan is based on the market and the companies we compete against for Team Members. The elements of our plan (e.g., base salary, bonus and stock options) are clearly similar to the elements used by many companies; however, our additional emphasis on fair treatment of all stakeholders requires that we cap executive and other leadership salaries at a level that does not prohibit us from competing for quality Team Members. The exact base pay, stock grant, and salary cap amounts are chosen in an attempt to balance our competing objectives of fairness to all stakeholders and attracting/retaining Team Members. EVA improvement is an objective calculation and was chosen as the basis for determining incentive compensation because we believe it is the best financial framework our executives can use to help make decisions that create sustainable shareholder value.

It is important to note that we have never lost an executive due to compensation. In addition, leadership turnover in the Company is less than 2% annually. We believe this is a good indication that our leadership compensation package is reasonable.

Subject to certain exceptions set forth below, Whole Foods Market plans stock option grant dates well in advance of any actual grant. Regarding our usual grants, the timing of each grant is determined months in advance to coincide with a scheduled meeting of our Board of Directors and its Compensation Committee. Except in highly unusual circumstances, we will not allow option grants at other dates. The grant date is established when the Company's Compensation Committee approves the grant and all key terms have been determined. The exercise price of each of our stock options grants is the market closing price on the grant date. The Company's general policy is for the annual grant to occur within two weeks after the official announcement of our second quarter results so that the stock option exercise price reflects a fully informed market price. This will usually be one week after the opening of the insider trading window. If at the time of any planned option grant date any member of our Board of Directors or Executive Team is aware of material non-public information, the Company would not generally make the planned stock option grant. In such event, as soon as practical after material information is made public, the Compensation Committee will have a specially called meeting and/or otherwise take all necessary steps to authorize the delayed stock option grant. Regarding the grant process, the Compensation Committee does not delegate any related function, and executives are not treated differently from other Team Members.

Accounting and Tax Considerations

Our stock option grant policies have been impacted by the implementation of SFAS No. 123R, which we adopted in the first quarter of fiscal year 2006. Under this accounting pronouncement, we are required to value unvested stock options granted prior to our adoption of SFAS 123 under the fair value method and expense those amounts in the income statement over the stock option's remaining vesting period. On September 22, 2005, we accelerated the vesting of all outstanding stock options except stock options held by members of our executive team and certain stock options held by our Team Members located in the United Kingdom. We incurred a $17.4 million pre-tax non-cash share-based compensation charge in the fourth quarter of fiscal year 2005 related to the accelerated vesting. Based on historical Team Member turnover rates and the Company's best estimate of future turnover rates, we recorded an additional $3.0 million pre-tax non-cash share-based compensation charge in the fourth quarter of fiscal year 2006 to adjust this estimate for actual experience. Our current intent is to limit the number of shares granted in any one year so that annual earnings per share dilution from share-based compensation expense will ramp up but not exceed 10% over time. We believe this strategy is best aligned with our stakeholder philosophy because it is intended to limit future earnings dilution from options while at the same time retains the broad-based stock option plan, which we believe is important to Team Member morale, our unique corporate culture and our success.

Part IV

Lessons We Have Learned

Part IV is the conclusion of our exploration of value(s)-based management. In this section we summarize what has been learned since the early days of VBM, along with what we ourselves have learned about CSR and its relationship to VBM. We discuss the results of archival academic studies that utilize publicly available information, as well as survey studies that communicate directly with the firms. We present our views of both the necessary ingredients for a successful V_sBM program and the way in which things go wrong in unsuccessful attempts. Finally, we conclude by looking at the current state of V_sBM and putting forward our vision of its future.

Chapter 9

Lessons Learned

> We are never confused about why we exist. Although volume growth, earnings, returns and cash flow are critical priorities, our people understand those measurements are all simply the means to the long-term end of creating value for our share owners.... I wrestle over how to build shareholder value from the time I get up in the morning to the time I go to bed. I even think about it when I'm shaving.
>
> —Roberto Goizueta, former chairman and CEO, Coca-Cola Company

Value-based management programs certainly appear quite simple on paper. Seemingly all that is needed to produce extraordinary wealth is to first adopt a VBM metric such as EVA and then to tie it to employee incentive compensation. Many companies appear to have reasoned that, if they merely hired an experienced consultant to implement the VBM program, the consultant could simply flip a magical switch, and the spigot of wealth would be opened.

Unfortunately, we now know the process is not nearly that simple. In fact, the initial wave of euphoric optimism has been replaced by some skepticism and resistance. We find this unfortunate and primarily the result of misunderstandings about the precise nature of a VBM program and what it takes for it to be successful. We further believe these are responsible for the misconception that VBM programs are in conflict with the desire to be socially responsible. Specifically, the unification of VBM and CSR (our value[s]-based management) forms a virtuous cycle of doing good and doing well as opposed to the wrong idea that VBM programs take wealth from nonshareholder stakeholders for the benefit of shareholders.

In this chapter we review the lessons that we have learned about the success and failure of VBM programs. To accomplish this we first review the findings of a variety of studies that examine the effects on company policies and performance that result from the adoption of a VBM program. We find a high degree of consistency among the studies supporting the old adage that "you get what you measure and reward." However, we find that the effectiveness of a VBM program is critically dependent on how the program is practiced. Overall, we observe that most of the early experiences have been positive, although the number of successful adoptions appears to have decreased over time and a number of firms have abandoned their VBM program.

Out of these research experiences we glean many important and consistent lessons. Specifically, successful VBM programs (1) have top-management support in the form of a strong commitment to value creation; (2) are tied to compensation; (3) entail a significant investment of time and money in educating the firm's workforce about the program; (4) are kept as simple as possible; and (5) focus broadly on the drivers of VBM rather than narrowly on the chosen metric.

VBM Studies Based on Archival Data

In one of the first studies that attempted to determine whether the claimed incentives of VBM metrics really work, we studied the actions of a sample of forty firms that had adopted compensation plans based on residual income, defined as earnings before interest less a capital charge on total capital (Wallace 1997). The actions of these firms were then compared to those of a matched pair of control firms that continued to base their incentive compensation on traditional accounting earnings (e.g., earnings per share or operating profits). The study concluded that it was indeed true that "you get what you measure and reward." When compared to the control sample, the VBM adopters (1) decreased their new investment and increased their dispositions of assets; (2) increased their payouts to shareholders through share repurchases; and (3) utilized their assets more intensively. All of these responses are consistent with value creation in which the VBM adopters and control firms each face similar investment opportunities. That is, all else remaining the same, disposing of nonproductive assets (i.e., those that do not produce a return equal to or greater than the firm's cost of capital), returning to the firm's stockholders any cash flow that is not needed to support the company's wealth-creating investment opportunities (i.e., dispensing free cash flow), and getting greater use out of existing assets are all ways to increase shareholder value. These actions are also consistent with a stronger awareness of the opportunity cost of capital, an incentive embedded in the use of a VBM metric tied to incentive compensation.

Balachandran (2006) extends the work of Wallace (1997) by examining the investment patterns of VBM adopters, depending on what type of prior incentives they were facing. Specifically, Balachandran found that switching from an earnings-based metric to a VBM metric decreased investment, which is consistent with our previous findings. An interesting result, however, is that firms that switched from a return on investment metric actually increased their investment. This result is consistent with what one would expect from VBM incentives and further supports the claim that VBM incentives work as claimed.

VBM Studies Based on Survey Data

In 1996 two of the authors of this book served on a research team sponsored by the American Productivity and Quality Center's International Benchmarking Clearinghouse, which conducted a study of businesses that had adopted and implemented a VBM system.[1] The objective of the study group was to document the industry's best practices in the use of VBM. Thus, an attempt was made to identify VBM success cases and then document their stories in depth.

A four-stage, data-collection process was used to document the case histories of VBM successes. A mail questionnaire was first sent to more than ninety potential candidates as a qualifying screen. We then phoned the respondents to further assess the capabilities of the their firms and determine their willingness to participate in the subsequent phases of the study. Next, the study group selected five benchmark partner firms for further study and site visits. These five companies represented a sampling of the VBM methodologies currently in use. Each one had not only adopted a VBM metric but had also integrated a complete process into the firm's evaluation structure. Three of the firms had adopted EVA; one used cash flow return on investment; and the fifth had selected from the components of a number of systems and designed its own program.

Factors Critical to the Success of VBM

Based on the findings of this study, we concluded that four factors were critical to a successful VBM program. As you might expect, these key factors are not unrelated to our basic findings discussed earlier. In fact, they come directly from the choices made in implementing and operating the system:

Lesson 1: Top-Management Support

The importance of management support cannot be over emphasized. In the words of one manager, "If a VBM program does not have the support of senior management, it is doomed to fail." Another said, "If the CEO doesn't buy into VBM, you are wasting your time."

The organization's top management must not only approve VBM but also be actively involved in promoting its use throughout the organization. All of the company representatives we interviewed demonstrated a commitment from top management. This effort creates a chain reaction that brings about the alignment of shareholder value with the corporate mission, vision, values, and strategic plan:

> Our use of EVA was largely a result of our CEO's vision of what the firm needed to do in order to become a more focused company. As a consequence, top-management support is communicated through the firm's vision and mission. Our mission statement links improvements in shareholder value to management's ability to fulfill its commitments to customers and create a work environment that maximizes employee performance.

Lesson 2: Performance and Compensation Are Tightly Linked

The VBM system gets its teeth from its connection to compensation that is based on employee performance. Without this connection, VBM becomes just another accounting or finance exercise:

> At our firm, the incentive compensation plan is the same for the chairman as it is for every salaried person in the office. Upon implementation of the compensation plan, there was an immediate behavioral/cultural change. For one thing, the quality of earnings improved; people were no longer trying to accomplish earnings by changing accruals. The budgets were no longer negotiated. There was a longer-term perspective on operations. Employees no

longer make decisions on a quarter-to-quarter basis, and they are held accountable for their decisions. The incentive plan encourages managers to think long term.

Lesson 3: Education Is Critical to Employee Acceptance

The executives of the different companies learned that, without an education process, their employees were not going to understand their role and that value enhancement would consequently be largely superficial. Otherwise, employees would not understand fully how their actions impact shareholder value. This requires a program of ongoing training for the employees so that they continually think about the impact of their actions on shareholder value: "Our managers devoted a full year for training. Then the compensation system went into effect the next year. In that way, the employees had time to get used to the program and become knowledgeable about it."

Lesson 4: Keep It Simple

In almost everyone's opinion, a firm should keep a VBM system as simple as possible. To be effective, the program must be understood and trusted by the employees. As one manager commented, "If the rank-and-file workers view the VBM program as simply a finance exercise, then the program is doomed." That is, genuinely understanding the VBM system is essential to its effectiveness. For that to happen, simplicity is a virtue—even at the expense of accuracy. However, most of the company representatives we interviewed believed that employees need to understand only the segment of the calculation for which they are responsible. Also, simplicity, according to some, requires consistency of measures throughout the firm; otherwise, there is only a limited buy-in:

> We determined very early on that we were not going to make the employee in the field an expert on EVA. He is not going to know the intricacies of the calculation; however, he is going to know the value drivers that he could impact. His performance is going to be judged on how well he does versus these drivers. It is our belief that understanding the concepts that can change management behavior is more important than understanding the technicalities of the actual calculation.
>
> One key to the success of VBM is that individuals are responsible for only the calculations they can control. They are not given large, wide-ranging directives like increased revenue, decreased expense, or limited capital spending. Management breaks the

calculation apart and sets goals and targets for employees on the specific aspects of VBM they control.

Managers simplified the financial-measurement framework to communicate information necessary to identify value drivers. Employees now understand how these drivers affect stock prices.

Managers developed measures that were the same for everybody. The capital definitions were the same for the manager as they were for the sweeper. A union representative made an interesting comment, "As long as you are computing capital the same way for me as you are for the CEO, then I've got no problem."

More Recent Survey Evidence

Around the same time as this research was being conducted, we also followed up on the findings of Wallace (1997) with a survey methodology (Wallace 1998). Like the archival study, this research also examined the management actions associated with a change to a compensation system based on residual income. We sent a questionnaire to a member of top management of seventy-six firms that have adopted EVA-type performance measures, forty of whom have included the measure in their incentive compensation. Based on the respondents' answers, we conclude that EVA performance measures appear to help align management's interests with those of the firm's shareholders and that the emphasis shifts from bottom-line earnings to earning more than the cost of employed capital. The responses, when compared with the prior empirical work, indicate a high degree of consistency among management actions attributed to the VBM incentives.

Consistent with the American Productivity and Quality Center's International Benchmarking Clearinghouse study, the responses in this study indicate that the respondents were pleased with their respective performance measures. As expected, the group of firms that tied the VBM metric to incentive compensation reported using the measure in more situations, obtained more changes as a result, and experienced more overall satisfaction with the measure than did those firms that did not tie it to incentive compensation. In addition, the responses to questions about the decisions and actions associated with the adoption of VBM metrics are consistent with the empirical findings in Wallace (1997).

We next describe a comprehensive survey of VBM firms that was conducted about five years later by a second team of researchers (Haspeslagh, Noda, and Boules 2001). By the time their research was undertaken, the earlier euphoria surrounding VBM had largely worn off, and many firms were beginning to question whether VBM was correct for them. The researchers sought to learn why many companies were reporting mediocre results or abandoning VBM programs while others were claiming great benefits from their VBM programs.

Haspeslagh, Noda, and Boules conducted a comprehensive survey of the use of VBM among the world's largest companies. Their survey resulted in 112 responses from leading VBM practitioners in North America, the United Kingdom, continental Europe, and Japan. In the researchers' words, "The picture of VBM that emerged turned out to be significantly different from the financial one commonly portrayed. For leading-edge practitioners, VBM is a holistic management approach that encompasses redefined goals, redesigned structures and systems, rejuvenated strategic and operational processes, and revamped human-resources practices. To these practitioners, VBM is not a quick fix but a path requiring persistence and commitment." The researchers reported that the greatest impact from the VBM programs was the employees' increased understanding of where value is created and destroyed and the managers' enhanced ability to make value-based decisions on issues such as allocation of resources.

The researchers noted that four ingredients were common to the successful VBM programs they studied. Notice that these four keys to success are nearly identical to those noted earlier in this chapter:

1. Companies must explicitly commit to shareholder value.
2. Extensive training is needed to ensure that VBM is thoroughly understood throughout the company.
3. The VBM measures should be linked to incentive compensation.
4. Systems and processes should be changed broadly rather than deeply, and the VBM program should be kept technically simple so that it is well understood.

While Haspeslagh et al. found many commonalities among the successful VBM adopters, they stressed that the most significant finding is that a successful VBM program is more about a cultural change within the company than it is about a financial tool. Brian Pitman, chairman of the British bank Lloyds TSB stated it well: "Anyone who thinks that implementing shareholder value is a matter of changing a few accounting systems and people will follow is fooling themselves" (66). The first challenge that companies faced when adopting VBM was to replace the prior mindset with one that included a commitment to value rather than to directives such as growth or accounting earnings.

Cadbury Schweppes chairman John Sunderland noted, "Managing for value is 20 percent about the numbers and 80 percent about people...because people create value" (68). The key to getting people to get on board the VBM program is extensive training. Haspeslagh et al. note that success is associated with education and that successful implementers invest heavily in it. They also state that education and training cover almost all managers and a significant percentage of nonmanagerial employees.

Successful VBM programs have almost always included raising the employees' stake in the company through incentive compensation. The adage that you get what you measure and reward, something we have repeated many times, is evident in the successful adoptions. Simply measuring something is rarely enough, as employees are far more likely to act in ways that enhance their rewards. As Sunderland has pointed out, the success or failure of the VBM program depends on how much it changes the way people behave. Haspeslagh et al. (2001) state that "VBM led to significant changes in behavior, for example, making employees more motivated and more accountable for their actions and expanding their skills sets."

A key finding of this research is that successful adopters realize that a VBM program is much more than just a metric. In fact, perhaps the biggest distinction between successful and unsuccessful VBM adoptions is the realization that VBM programs are about a cultural change within a company and that the VBM metric represents only a scorecard. Haspeslagh et al. note that most unsuccessful VBM companies focus their efforts almost entirely on changing their accounting systems and developing complex performance measures. Successful VBM companies, in contrast, develop a far more comprehensive view of their company's processes. Rather than simply promoting a particular metric, they identify its value drivers. By focusing on these value drivers they are able to help frontline employees understand how they can create value:

> The benefits of VBM can be significant if implementation is done right, but the view still prevalent in some quarters that VBM is a miracle cure with near-instantaneous effect turns out to be unrealistic. Companies contemplating adopting VBM must be prepared to stick with it for many years and must not expect too much in the early going.... VBM, applied properly as a holistic agent of change, will solve a great many of your problems and will put your company's profitability back on track. At the very least, it will build a profitable springboard from which you can leap into the next phase of your climb toward sustained shareholder-value creation.

Current CSR Research

The recent upsurge in interest surrounding CSR piqued our own interest and led us to conduct a wide-scale research project to address, among other questions, the ways in which CSR affects a firm's financial performance and its shareholder wealth creation (Lougee and Wallace 2008).

Our study utilized the extensive database KLD STATS provided by KLD Research and Analytics to document CSR activity over the past fifteen years for both the Standard and Poor's 500 firms and the Domini 400 firms.[2] The database includes more than ninety social and environmental indicators in seven broad categories: community, corporate governance, diversity, employees, environment, human rights, and products. Each category contains a series of potential strengths (e.g., charitable giving under "community") and concerns (e.g., hazardous waste under "environment"). We merged these data with financial data for each of these firms.

As we have already noted, the existing academic literature has found support for the premise that CSR makes financial sense. For the most part, these studies looked at how well investments in firms with high CSR have fared relative to benchmark portfolios such as the S&P 500. We noted similar evidence that documents that the Domini 400 index has slightly outperformed in benchmark S&P 500 firms for the entire period from its inception in April 1990 through December 2006. The Domini 400 index has yielded an annual rate of 12.09% for this period, compared to an annual rate of 11.45% for the benchmark S&P 500.

While it is one thing to show that socially responsible investing (SRI) can make financial sense, this is not the same thing as showing that it makes good business sense for firms to invest in CSR programs. In order to test this latter question, we looked at how CSR relates to the performance of individual firms. We employed regression analysis to study the connection between a firm's CSR strengths and concerns and its financial performance as measured by return on assets (ROA).

A company's CSR strengths showed a highly significant positive association with the firm's ROA, whereas its CSR concerns demonstrated a highly significant negative association with the firm's ROA. This same relationship was exhibited for both the Domini 400 and the S&P 500 firms. We further found that this relationship is generally consistent among the different categories of CSR strengths and concerns. Overall, both the prior academic evidence and our separate test of CSR and firm financial performance tell a consistent story: CSR programs make sense from a financial perspective.

Summary

It now seems that sufficient time has passed and enough lessons have been learned that we can identify many commonalities in the experiences of VBM firms. These common attributes are found in numerous independent studies that have employed a variety of methodologies. The following points are clear:

1. Value-based management is not a quick-fix panacea that can be implemented without serious effort.
2. To be successful, VBM requires a cultural change within a company. Simply picking a VBM metric is far from sufficient.
3. A sincere commitment to a culture of value creation is essential. This commitment is usually initiated from the top.
4. Extensive training is required so that everyone understands the VBM program and the ways in which they individually and as a team can create value. While extensive education is an attribute of successful programs, so is keeping things as simple as possible.
5. In order to change behavior, firms must both measure and reward. Successful VBM programs are characterized by a tight integration of VBM measures and incentive compensation in order to inspire employees to think and act like owners.
6. By creating a virtuous cycle of doing good and doing well, V_sBM, the combination of VBM and CSR into a value(s)-based management approach, can lead to a win-win situation for the firm's shareholders and its numerous other stakeholders.

If used effectively, VBM is more than a financial exercise; its objective is also a cultural change that involves aligning the interests of a firm's management with those of its shareholders. We found that a company that successfully integrates VBM throughout its organization experiences a dramatic change in its culture. Certainly not everything has worked perfectly, and tradeoffs invariably had to be made between accuracy and simplicity; moreover, some companies also experienced significant resistance to such major changes. However, for successful VBM companies, the rewards far outweighed the costs.

Like VMB, CSR also has as its objective a cultural change in the mindset of every employee. In other words, CSR promotes the idea that firms must operate in a socially responsible manner such that they consider how their operations impact their numerous stakeholders. Evidence is strong that a company's many stakeholders are critical ingredients for its long-term sustainable value creation.

Finally, none of the companies we studied viewed the choice of the metric itself as the key to success. While the selection process in each case no doubt considered finding the best fit between the metric and a company's characteristics, the ultimate key to success was effective implementation. Myers (1997) has put this fact in terms that both golfers and nongolfers can understand: "Much as hitting a good golf shot depends more on how you strike the ball than on which brand of club you use, achieving success through the use of any performance metric will depend more on how well you apply it than which one you use."

So let there be no question—managing for value will always be a challenge, even in the best of circumstances. Thus, we should always be a bit suspect of anyone who tries to convince us that managing for value is easy if we will only do it their way. Also, we should never believe that simply adopting a system is what counts; what matters instead is how well we implement it. From what we have observed, first choosing a method that management can believe in and support and then becoming determined to be the best at implementing it will take us a long way down the road to success. And in listening to those managers who have been successful in this endeavor, the journey is well worth the effort.

Epilogue: Where We Are Now

We sincerely hope that you have learned as much from reading this book as we have learned from writing it. We had initially believed that our first book, *Value Based Management: The Corporate Response to the Shareholder Revolution*, published in 2000, still captured the basic ingredients of VBM and just needed some updating. What we soon realized, however, is that things have changed markedly in that short time. It is not that the fundamentals of VBM are any different. Quite the contrary, we firmly believe they are just as relevant today (and perhaps even more so) in the increasingly competitive global market. What has changed—and in no subtle way—is the perception of wealth creation and VBM.

What struck us early on in our research for this book was the dearth of VBM articles in the popular press. Also absent were companies' testimonials about their VBM experiences. These were in no short supply a few years ago, when we were writing our first book. Naturally intrigued, we set about learning why the current state of affairs exists.

So what did we find out? While it is difficult to ascertain the exact reasons for the present situation, we believe that several factors are at work. One results from the tremendous level of excitement and fanfare that initially surrounded these new VBM programs. As it turns out, VBM was likely oversold and thus unable to live up to the unreasonable expectations that companies had formed. As we note in chapter 9, VBM is not a quick-fix panacea that can be implemented without serious effort. To be successful, it requires a cultural change. Simply picking a VBM metric is far from sufficient. Too many firms have been unwilling to make the required commitment and have quickly become disillusioned.

While the early successes and euphoria surrounding VBM may have contributed to the current situation, we believe that another, far stronger cause is at work. Unfortunately, misinformation and misunderstanding are largely to blame for what we believe is the primary reason that fewer firms are embarking upon VBM programs today. Value-based management has likely become another victim of the large accounting scandals of the past few years. No firm wants to be thought of as another Enron. Moreover, many people incorrectly believe that a shareholder wealth focus was largely to blame for what happened at firms like Enron.

A closer inspection of the Enron story will prove that the complete opposite is the case. Not only was VBM not at fault at Enron, but a VBM program would likely have prevented the series of events that led to this horrendous scandal. Bennett Stewart, a cofounder of Stern Stewart & Co. and a leading proponent of VBM, set forth a convincing argument in his article "The Real Reasons Enron Failed" (2006). In that article Stewart notes that a combination of errors from *not* following a VBM paradigm resulted in the Enron failure.

Many people attribute the fraudulent accounting practiced at Enron for its eventual demise. Rather than causing the downfall, however, it appears the accounting was simply a consequence of other factors and was initiated in an attempt to keep the firm from failing. Not that the accounting practiced at Enron was not to blame, however. It is just that the blame should be placed on the accounting, which was considered to be based on generally accepted accounting principles (GAAP) rather than the dishonest financial practices we have heard so much about.

Somehow Jeff Skilling was able to persuade the Securities and Exchange Commission and the auditing firm Arthur Anderson that mark-to-mark accounting (M2M) was an acceptable way to record revenue. In actuality, M2M allowed Enron to record revenue at the time a contract was signed rather than ratably over the life of the contract, which may be ten to twenty years. It is, of course, quite understandable that Enron management would want an accounting procedure that greatly accelerates earnings since the company's annual bonuses were calculated as a percentage of reported net income. This focus on earnings per share (EPS), based on accounting net income, was no secret, as was highlighted in Enron's letter to shareholders in 2000: "Enron is laser-focused on EPS, and we expect to continue strong earnings performance" (quoted in Stewart 2006, 116–17).

The trouble is that these strong earnings, while greatly enriching Enron's management, had little connection to economic value. Enron CFO Andrew Fastow realized this, but that did not matter, as the following policy declaration attests: "Reported earnings follow rules and principles of accounting. The results do not always create measures consistent with underlying economics. However, corporate management's performance is generally measured by accounting income, not underlying economics. Risk management strategies are therefore directed at accounting rather than economic performance" (quoted in Stewart 2006, 119).

We doubt there is any clearer example of the adage "you get what you measure and reward." Even though Enron management knew the strategy was incorrect, the company's executives were willing to make operating and risk-management decisions in order to increase their bonuses even with the clear knowledge that these actions would destroy economic value. After all, this is what they were being paid to do.

Furthermore, M2M accounting was not the only problematic accounting practice at Enron. A far larger distortion resulted from accounting's failure to deduct a cost for shareholders' capital. This omission allowed accounting earnings to vastly overstate the economic earnings that would have been calculated under a VBM measure such as EVA. Enron was driven to make deal after deal, which resulted in increased earnings even though these transactions failed to cover their cost of capital. Again, this is what management was being measured on and rewarded to do. Eventually Wall Street caught on to these low returns, and this discovery finally caused the house of cards to come tumbling down. Had Enron been run with a VBM discipline and been measured and rewarded on the basis of economic wealth creation rather than growing accounting earnings, it is doubtful that it would ever have needed to resort to the accounting trickery that has been its most visible legacy.

Stewart's well-reasoned argument does not change the perception held by much of the public that the cause of Enron's failure was both greed and a "laser focus" on shareholder value rather than a "razor focus" on EPS. This widespread misperception, coupled with similar scandals at other large corporations such as WorldCom, led to the Sarbanes Oxley (SOX) set of governmental regulations. The last thing a leading corporation wants in this environment is to be labeled as greedy and socially irresponsible. We believe that the resulting elevation of the corporate social responsibility movement within corporate boardrooms is largely an outgrowth of the fear that board members have of being associated with Enron-like behavior.

Stewart provided us with this explanation of much of the lack of demand for VBM: "I think Sarbanes Oxley had a chilling effect on value-based management. It made boards ultra conservative and encouraged them to hew ever more closely to accounting and to be far less open to non-GAAP measures like EVA."[1]

The rise of CSR and the reluctance to be explicitly associated with VBM have been trademarks of European corporation for many years. These viewpoints are natural outgrowths of socialistic-leaning governments and very strong labor unions. Against this backdrop, Stern, Shiely, and Ross (2001) discuss a study by Pascal Luciani, written when he was an MBA candidate at the London Business School, titled "EVA in Europe—A Cultural Perspective." Luciani's research was based on extensive interviews with business people in several European countries. He concluded that EVA faces major obstacles in Europe because companies there take social responsibility very seriously. Moreover, he learned that these companies found it easier to talk about the interests of stakeholders than those of shareholders.

This same sentiment was also noted in the study by Haspeslagh, Noda, and Boules (2001) we discuss in chapter 9. These authors note that "in some countries, the very term 'shareholder value' is considered politically incorrect" (67). They give an example of a CEO of a large multinational that prominently portrayed the company's VBM program in the United States but downplayed it in France. "I drive differently in the U.S. than I do in France. I also don't manage in the same way" (67–68). The CEO was convinced of the merits of openly espousing shareholder value; however, his French board members worried that such a forthright commitment in France would antagonize the country's government and unions.

So the question becomes whether the recent response to VBM marks its end as a driving force behind wealth creation or is merely a temporary reaction to an unfortunate set of events. We certainly believe that it is only a matter of time before VBM returns to the forefront of corporate agendas. Furthermore, Peter Drucker's words are undoubtedly as relevant today as they were when he wrote them more than ten years ago: "A business that does not show a profit at least equal to its cost of capital is socially irresponsible; it wastes society's resources. Economic profit performance is the base without which business cannot discharge any other responsibilities, cannot be a good employer, a good citizen, a good neighbor" (2002b, 84).

The purpose of the corporation (discussed at length in chapter 1) continues to be the creation of wealth. Only then is society as a whole made better since only then will wealth be created to satisfy the needs of the many corporate stakeholders. As Drucker notes, it is socially irresponsible to manage in any other way.

Without a doubt, it appears that many CEOs, CFOs, and boards became very focused on compliance with SOX at the expense of operations. One of the authors of this book recently attended a Harvard conference for audit chairs. The sense of the group members was that things are better, but they are still irritated by SOX and its costs. However, much of the pain is now behind them, and they are returning to pre-SOX priorities. From the board's perspective, performance measurement is once again becoming a real issue. Clearly the need to measure and reward performance is a problem that never goes away, and value(s)-based management is arguably the best means to accomplish these tasks.

We hope that this book contributes to the education needed to clear up the misconceptions about the relationship between VBM and CSR. In their practice of value(s)-based management, Whole Foods Market and Herman Miller are living proof that the two concepts are intertwined. They also admirably demonstrate that explicitly affirming both VBM and CSR is far from inconsistent. Rather, these philosophies are strong complements of one another.

Notes

Preface

1. The American Productivity and Quality Center was founded in 1977 as a nonprofit, 501(c)(3) organization. Its mission is to improve productivity and quality in the private and public sectors. The study was titled "Shareholder Value-based Management."

Chapter 1

1. We use the term *value(s)-based management* to refer to performance metrics and reward systems that are designed to help managers enhance long-term firm value.
2. This example, based on Michael Jensen's 2001 article "Value Maximization, Stakeholder Theory, and the Corporate Objective Function," assumes, for simplicity, that profit streams are level and perpetual. In the more realistic situation of uneven cash flows over time, the firm would need to consider present values; however, the basic concept remains the same.

Chapter 2

1. To draw an analogy, VBM has more in common with steering supertankers than it does with guiding missiles by laser. The idea is to point the firm in the direction that offers increased likelihood of creating value rather than trying to specify value creation to the second decimal place.

2. If you are still unconvinced, consider the following example. In 1999 Intel Corporation spent $3,503 million on R&D, which equals $0.69 per share. In addition, the firm's 1999 earnings were $2.20 per share. On March 10, 2000, the firm's stock sold for $135.00, producing a price/earnings multiple of 61.36 times. Ask yourself the following question: Do you believe Intel's stock price would have been $42.13 ($0.69 in R&D per share × 61.36 times) higher if Intel had not spent anything on R&D? Of course not—because Intel's expenditures for R&D are its life's blood in creating new products that drive its future profitability. Without its R&D the firm's stream of new products would evaporate, as would its future cash flows.

Chapter 3

1. Stern Stewart and Co., "EVA and the Balanced Scorecard" (1998).
2. Stern Stewart Europe Limited, "ABC, the Balanced Scorecard, and EVA" (1999).
3. Total shareholder return is defined as price appreciation plus dividends $(price_{end} - price_{begin} + dividends)\ /\ price_{begin} \ldots$
4. For the time being we simply use the term *economic profit* rather than one of the VBM metrics we study in part II of this book. Economic profit represents the same concept that each of the VBM metrics attempts to measure.
5. The project has an annual economic profit of $25,000 in perpetuity, which, given the firm's 10% cost of capital, has a present value of $250,000, which equals the project's NPV. The consistency of EVA with the NPV rule in this instance is not always the case. For a discussion of the limitations of EVA in this regard see Martin, Petty, and Rich (2005).
6. Interest expense of 6% on one-third of the total investment requires that the project earn only 2% on the total investment.

Chapter 4

1. See Rappaport (1998).
2. See the bible of EVA, *The Quest for Value,* by G. Bennett Stewart III (1991), and a more recent presentation by Joel Stern, John Shiely, and Irwin Ross, *The EVA Challenge* (2001).
3. Non-interest-bearing current liabilities (NIBCLs) are part of a firm's operating cycle as it purchases inventory on credit, sells on credit, pays its accounts payable and its accrual accounts, and eventually collects from its customers. Interest-bearing debt, on the other hand, is a source of financing provided by return-seeking investors and, as such, is not part of the firm's recurring operating cycle.
4. Judgment is not necessarily a bad thing within the computation of earnings. Certainly, there is potential for bias if the managers are so inclined,

and this can lead to a loss of credibility. However, earnings are possibly better at forecasting future earnings because the accrual system brings the future into the present, which, of course, involves judgment—and potential bias.

5. The reason that this equation is the present value of a future cash-flow stream that continues in perpetuity is not apparent. However, it is true and is demonstrated in most basic financial textbooks, so we will leave the derivation to others and simply rely on their proofs.
6. For instance, the projected free cash flow in year 1 is $4,125; its present value is $3.619 million, computed as follows:

$$\begin{array}{l}\text{present value of}\\ \text{year 1 free cash flow}\end{array} = \frac{\text{free cash flow in year 1}}{(1+\text{cost of capital})^1} = \frac{\$4.125}{(1+.14)^1} = \$3.619 \text{ million.}$$

7. To learn the intricacies of computing a company's cost of capital, see Titman and Martin (2007).
8. The market risk premium is the rate of return above the risk-free rate that an investor would be expected to earn on a well-diversified portfolio of stocks (Ibbotson and Singuefield 2007).

Chapter 5

1. The acronym EVA is registered by Stern Stewart and Co.
2. Cash flow return on investment, as promoted by Boston Consulting Group, was once a contender but is no longer. The company is even reluctant to acknowledge any affiliation with CFROI as a tool for corporate management. The metric is, however, still used by Credit Suisse First Boston Holt as a tool for money managers. Given its historical presence, we have provided a brief description of the technique in appendix 5C.
3. We are assuming that the firm will continue to make *replacement* investments in the amount of the depreciation being taken, but it will not need to increase the total invested capital after 2013.
4. Not all of the adjustments that accountants make are unnecessary. For Stern and Stewart, depreciation is one noncash charge that is acceptable. Their rationale is that the assets consumed in the business must be replenished before investors achieve a return on their investment. Thus, depreciation is recognized as an economic cost, and capital should accordingly be charged with the accumulated depreciation suffered by the assets.
5. EVA Seminar, New York, November 2000.
6. http://www.sternstwert.com (accessed November 30, 2006).
7. For simplicity's sake, this explanation of residual income valuation is based on an all-equity firm. That is, profits are after-tax net income (not net operating profits after tax), and the opportunity cost of capital is the cost of equity (not the weighted average cost of capital).

Chapter 6

1. See De Ramos (2003).
2. Sir Howard Stringer is Sony's CEO.

Chapter 7

1. The $ROIC_t$ estimates are based on the use of present-value depreciation in determining operating income and the invested capital for the period.

Chapter 8

1. Simple examples of nonmonetary rewards that are important motivators of individual behavior within an organization include participation in the firm's activities, a safe and comfortable work environment, public recognition, satisfying interactions with fellow employees, and promotions. All of these are important elements of the firm's overall compensation system.
2. For example, the Boise Cascade Corporation (2006 10K annual report, Mar. 1, 2007, 126) states that its executive compensation program is designed to do the following:

   - closely align compensation with the performance of the company on both a short-term and a long-term basis;
   - link each executive officer's compensation to his or her performance and the areas of the company for which he or she is responsible;
   - attract, motivate, reward, and retain the broad-based management talent critical to achieving the company's business goals; and
   - encourage executive officers to own company management equity units.

3. Wruck (2000) suggests that effectively designed and implemented compensation systems create value for organizations by (1) improving employees' motivation and productivity, (2) promoting the productive turnover of personnel, (3) mobilizing valuable, specific knowledge by allowing effective decentralization, and (4) helping overcome organizational inertia and opposition to change.
4. Ferris and Wallace (2007) summarize the tremendous growth in CEO and top executive compensation for the period 1992–1994 and explain that it has involved a much higher proportion of at-risk or variable pay. From 1992 to 2004, total CEO compensation at S&P 500 firms rose by more than 200%, while salary remained nearly flat. Although total compensation rose by a little more than 200%, the really big increases in CEO compensation came in the form of grants of stock and options that rose by nearly 300%.
5. Much of this discussion is based on the work of G. Bennett Stewart III (2002)

6. Apparently the Financial Accounting Standards Board (FASB) has finally decided to follow Warren Buffet's lead. Stock option grants are now required to be expensed on the company's income statement as dictated under FASB 123R.
7. Bonus plans based on the EVA metric have evolved over time, and recent versions place more emphasis on EVA improvement. Early versions simply based the bonus on a percentage of EVA: Bonus = x% * EVA. The next generation of EVA-based bonuses was structured to have the bonus include a percentage of the level of EVA as before, plus a percentage of the change in EVA with the latter component typically the larger share of the bonus. The general form of these bonuses took the form: Bonus = (x% * EVA) + (y% * change in EVA). Recent versions of these plans are now in the general form Bonus = Target Bonus + y% * (change in EVA—expected improvement in EVA). Including expected improvement in the bonus formula is meant to incorporate the fact that firm value reflects both assets in place and the value of future growth opportunities, thus providing a better alignment between those receiving a bonus and the returns to investors (Desai and Ferri 2006).
8. Our discussion of the components of compensation policy follows that of Baker, Jensen, and Murphy (1988).
9. However, pay for performance is not accepted in certain quarters. Some psychologists and behaviorists have argued that pay for performance- or merit-based compensation can create dysfunctional results. Kohn (1988), for example, argues that merit-pay systems are counterproductive for three reasons: (1) "rewards encourage people to focus narrowly on a task, to do it as quickly as possible, and to take few risks"; (2) "extrinsic rewards can erode intrinsic interest"; and (3) "people come to see themselves as being controlled by a reward." Similarly, Deci (1972) argues that money actually lowers employees' motivation by reducing the "intrinsic rewards" that they receive from the job.
10. The following discussion is very basic. In most firms incentive pay is determined by multiple performance metrics. For example, executives may receive 75% of their incentive pay based on financial results and 25% based on personal objectives. In other cases it might be determined by individual, financial, and strategic performance.
11. Incentive pay = ($50,000 × .20) × 1.10 = $11,000, and total compensation = $50,000 + $11,000 = $61,000.
12. Implicit in our discussion is the assumption that the performance metric is bounded from below at zero.
13. Art Knight, CEO of Morgan Products, describes the problem as follows: "It used to be that our compensation plans for senior and divisional managers were tied, as so many plans are, to negotiated budgets—not to the enhancement of shareholder value. While meeting budget figures does enhance shareholder value, it may not give managers incentive to enhance value even more by going beyond budgeted numbers" (Edwards 1993).

14. We use the firm's 10% required rate of return to discount the managerial bonuses since these are contingent on the performance of the project. In addition, we set the pay for performance bonus equal to 1% of EVA for the period.
15. The manager does not have to leave the firm for the managerial horizon problem to arise. He may simply move to another operating unit or division where the performance of his current division does not impact his incentive compensation.
16. When asked whether there was a danger that managers would make shortsighted investment decisions to boost their EVA bonuses, CEO Randall Tobias of Eli Lilly responded, "Yes. One of the things you never want to set up is a system where there's a great incentive to do something stupid in the short term. Theoretically you could lower your asset base and improve your EVA by depleting all your inventory. But then, come the first of next year, you'd have nothing to sell.... To make sure that kind of shortsighted decision making doesn't happen, we've set up something called a bonus bank that encourages managers to take a longer-term perspective" (Martin 1996).
17. Sometimes the bonus-bank concept can be used in combination with another measure. For instance, if revenue growth is important (but secondary to the primary measure), a compensation system might be designed in which a portion of the earned bonus (on a sliding scale) is banked if revenue growth goals are not attained.

Chapter 9

1. The clearinghouse is located in Houston, Texas. The center was founded in 1977 as a nonprofit, 501(c)3 organization whose mission is to improve productivity and quality in the private and public sectors. The center has a staff of approximately one hundred, a governing board of directors, and a budget of around $16 million. The Benchmarking Clearinghouse organizes and performs benchmarking studies whereby groups of independent corporations are brought together to share the costs and benefits of their benchmarking efforts.
2. The Domini 400 is one of the portfolios of firms developed to serve as an index for socially conscious investing.

Epilogue

1. Stewart, speaking on a panel at the Ethics Forum, Baylor University, Waco, Texas, 2002.

References

Bagnoli, M., and S. Watts. 2003. "Selling to socially responsible consumers: Competition and the private provision of public goods." *Journal of Economics and Management Strategy* 12:419–45.

Baker, George P., Michael C. Jensen, and Kevin J. Murphy. 1988. "Compensation and incentives: Practice vs. theory." *Journal of Finance* 43(3):593–616. Repr., Michael C. Jensen, *Foundations of Organizational Strategy* (Cambridge, Mass.: Harvard University Press, 1998).

Balachandran, Sudhakar V. 2006. "How does residual income affect investment? The role of prior performance measures." *Management Science* 53(3):383–94.

Baron, D. 2001. "Private politics, corporate social responsibility, and integrated strategy." *Journal of Economics and Management Strategy* 10:7–45.

Berle, Adolph, and Gardner Means. 1932. *The modern corporation and private property.* New York: Macmillan.

Bierman, Harold. 1988. "Beyond cash flow ROI." *Midland Corporate Finance Journal* 5(4) (Winter):36–39.

Collins, James, and Jerry Porras. 2002. *Built to last: Successful habits of visionary companies.* New York: HarperCollins.

Collins, Jim. 2001. *Good to great: Why some companies make the leap... and others don't.* New York: HarperBusiness.

Davies, Erin. 1997. "What's right about corporate cash flow: Afloat in a sea of green." *Fortune* (March 31):28.

De Ramos, Abe. 2003. "Watch out, Sony: How Samsung's finance chief is taking the Korean electronics giant to the world." http://www.CFOAsia.com (April).

Deci, Edward. 1972. "The effects of contingent and non-contingent rewards and controls on intrinsic motivation." *Organizational Behavior and Human Performance* 8:217–29.

Degeorge, François, Jayendu Patel, and Richard Zeckhauser. 1999. "Earnings management to exceed thresholds." *Journal of Business* 72(1):1–33.

Desai, Mihir A., and Fabrizio Ferri. 2006. "Understanding economic value added." *Harvard Business School* (July 11):9–12, note 206-016.

Dixit, Avinash K., and Robert S. Pyndick. 1994. *Investment under uncertainty.* Princeton, N.J.: Princeton University Press.

Drucker, Peter F. 1954. *The practice of management.* New York: HarperCollins. Repr. 1993.

———. 1995. "The information executives truly need." *Harvard Business Review* (January/February):73.

———. 2002a. *A functioning society: Selections from sixty-five years of writing on community, society, and polity.* New Brunswick, N.J.: Transaction.

———. 2002b. *Managing in a time of great change.* Oxford, UK: Butterworth-Heinemann.

———. 2008. *Management.* Rev. ed. New York: Collins Business.

Drucker, Peter F., and Joseph A. Maciariello. 2004. *The daily Drucker: 366 days of insight and motivation for getting the right things done.* New York: HarperBusiness.

Edwards, Laurie. 1993. "You can't beat cash." *Across the Board* 30(7):20–22.

Ehrbar, Al. 1998. *EVA: The real key to creating wealth.* New York: Wiley and Sons.

Elliott, Lisa. 1997. "Is EVA for everyone?" *Oil and Gas Investor* 17(2):46–51.

Ferris, Kenneth R., and James S. Wallace. 2009. "IRC section 162(m) and the law of unintended consequences." Claremont Graduate University Working Paper.

Finegan, Patrick T. 1989. "Financial incentives resolve the shareholder-value puzzle." *Corporate Cashflow* (October):27–32.

Freeman, Edward. 1984. *Strategic management: A stakeholder perspective.* Boston: Pitman.

Friedman, Milton. 1962. *Capitalism and freedom.* Chicago: University of Chicago Press.

Gaver, Jennifer J., Kenneth M. Gaver, and Jeffrey R. Austin. 1995. "Additional evidence on bonus plans and income management." *Journal of Accounting and Economics* 19:3–28.

Goodyear, C. W. 2006. "Social responsibility has a dollar value." http://www.theage.com.au (July 27).

Hall, Brian J., and Kevin J. Murphy. 2000. "Optimal exercise prices for risk averse executives." *American Economic Review* (May):209–14.

———. 2002. "Stock options for undiversified executives." *Journal of Accounting and Economics* 33:3–42.

Handy, Charles. 2002. "What's a business for?" *Harvard Business Review* (December):54.

Haspeslagh, Philippe, Toma Noda, and Fares Boules. 2001. "Managing for value: It's not just about the numbers." *Harvard Business Review OnPoint* 79(7) (July–August):64–73, 144.

Hawawini, Gabriel, and Claude Viallet. 1999. *Finance for executives: Managing for value creation.* Cincinnati: South Western College Publishing.

Healy, Paul M. 1985. "The effect of bonus schemes on accounting decisions." *Journal of Accounting and Economics* 7:85–107.

Holthausen, Robert W., David F. Larcker, and Richard G. Sloan. 1995. "Annual bonus schemes and the manipulation of earnings." *Journal of Accounting and Economics* 19:29–74.

Ibbotson, Roger G., and Rex A. Singuefield. 2007. *Stocks, bonds, bills, and inflation: Historical returns.* Chicago: Ibbotson.

Institute of Management Accountants. 2007. "Executives say corporate responsibility can be profitable." *IMA Online Newsletter* (October 15).

Jensen, Michael C. 1998. *Foundations of organizational strategy.* Cambridge, Mass.: Harvard University Press.

———. 2001. "Value maximization, stakeholder theory, and the corporate objective function." *Bank of America Journal of Applied Corporate Finance* 14(3) (Fall):8–21.

Jensen, Michael, and William Meckling. 1998. "Divisional performance measurement." In Michael C. Jensen, *Foundations of Organizational Strategy,* chapter 12. Cambridge, Mass.: Harvard University Press.

Kohn, Alfie. 1988. "Incentives can be bad for business." *Inc.* (January):93–94.

Lougee, Barbara, and James S. Wallace. 2008. "What the data tell us about the corporate social responsibility (CSR) trend." *Journal of Applied Corporate Finance* 20 (1):96–108.

Lucier, Chuck. 2004. "Herb Kelleher: The thought leader interview." *strategy+business* (Summer).

Martin, John D., and J. William Petty. 2000. *Value-based management: The corporate response to the shareholder revolution.* Boston: Harvard Business School Press.

Martin, John, J. William Petty, and Steve Rich. 2005. "A survey of EVA and other residual income models of firm performance." *Journal of Finance Literature* 1:1–20.

Martin, Justin. 1996. "Eli Lilly is making shareholders rich. How? By linking pay to EVA." *Fortune* (September 9).

Mauboussin, Michael J. 1995. "Wealth maximization should be management's prime goal." Credit Suisse First Boston, Equity Research-Americas (December 13).

McCormack, John, and Jawanth Vytheeswaran. 1998. "How to use EVA in the oil and gas industry." *Journal of Applied Corporate Finance* 11 (Fall):109–131.

McTaggart, James M., Peter W. Kontes, and Michael C. Mankins. 1994. *The value imperative: Managing for superior shareholder returns.* New York: Free Press.

McWilliams, A., and D. Siegel. 2000. "Corporate social responsibility and financial performance: Correlation or misspecification?" *Strategic Management Journal* 21:603–609.

Murphy, Kevin. 1999. "Executive compensation." In Orley Ashenfelter and David Card, eds., *Handbook of Labor Economics,* vol. 3. Amsterdam: North Holland.

Myers, Randy. 1997. "Measure for measure." *CFO Magazine* 13 (November):44–56.

O'Byrne, Stephen F. 2000. "Does value-based management discourage investment in intangibles?" In James L. Grant and Frank J. Fabozzi, eds., *Value-based metrics: Foundations and practice.* New Hope, Penn.: Frank J. Fabozzi Associates.

Orlitzky, M., F. Schmidt, and S. Rynes. 2003. "Corporate social and financial performance: A meta-analysis." *Organization Studies* 24(3):443–41.

Porter, Michael E., and Mark R. Kramer. 2006. "Strategy and society: The link between competitive advantage and corporate social responsibility." *Harvard Business Review OnPoint* (December).

Rappaport, Alfred. 1998. *Creating shareholder value: A guide for managers and investors,* 2d ed. New York: Free Press.

"Rethinking the social responsibility of business: A *Reason* debate featuring Milton Friedman, Whole Foods' John Mackey, and Cypress Semiconductor's T. J. Rodgers." 2005. *Reason Magazine* (October).

Sanders, W. Gerald, and Donald Hambrick. 2007. "Swinging for the fences: The effects of CEO stock options on company risk-taking and performance." *Academy of Management Journal* 50 (October/November):1055–78.

Siegel, D., and D. Vitaliano. 2006. "An empirical analysis of the strategic use of corporate social responsibility." Working paper. Available at Social Science Research Network (SSRN), http://ssrn.com/abstract=900521 (April).

Smith, Adam. 1776. *The wealth of nations.* Repr., London, Penguin, 1999.

Stern, Joel M., John S. Shiely, and Irwin Ross. 2001. *The EVA challenge: Implementing value-added change in an organization.* New York: Wiley and Sons.

Stern Stewart and Co. 1998. "EVA and the balanced scorecard." *EVAngelist* 2(4).

Stern Stewart Europe Limited. 1999. "ABC, the balanced scorecard, and EVA." *EVAluation* 1(2) (April).

Stewart, G. Bennett, III. 1991. *The quest for value.* New York: HarperBusiness.

———. 2002. "How to structure incentive plans that work." *EVAluation* 4(4).

———. 2006. "The real reasons Enron failed." *Journal of Applied Corporate Finance* 18(2) (Spring):116–19.

Titman, Sheridan, Arthur Keown, and John Martin. Forthcoming. *Principles of finance.* New York: Prentice Hall.

Titman, Sheridan, and John Martin. 2007. *Valuation: The art and science of making strategic investments.* Boston: Pearson Addison Wesley.

Trigeorgis, Lenos. 1996. *Real options: Managerial flexibility and strategy in resource allocation*. Cambridge, Mass.: MIT Press.

Tully, Shawn. 1993. "The real key to creating wealth." *Fortune* (September 20):38–50.

———. 1998. "America's greatest wealth creators." *Fortune* (November 9):193–204.

Wallace, James S. 1997. "Adopting residual income-based compensation plans: Do you get what you pay for?" *Journal of Accounting and Economics* 24 (December):275–300.

———. 1998. "EVA financial systems: Management perspective." *Advances in Management Accounting* 6:1–15.

Welch, Jack, and John A. Bryne. 2001. *Jack: Straight from the gut*. New York: Warner Business Books.

Williamson, Robert M. 2006. "What gets measured gets done: Are you measuring what really matters?" http://www.swspitcrew.com/articles/What%20Gets%20Measured%201106.pdf (accessed December 24, 2008).

Wruck, Karen Hopper. 2000. "Compensation, incentives, and organizational change: Ideas and evidence from theory and practice." In Michael Beer and Nitin Nohria, eds., *Breaking the code of change*. Boston: Harvard Business School Press.

Young, S. David, and Stephen F. O'Byrne. 2001. *EVA and value-based management*. New York: McGraw-Hill.

Zimmerman, Jerald L. 1997. "EVA and divisional performance measurement." *Journal of Applied Corporate Finance* 10:98–109.

Zimmerman, Ross L. 2006. *Accounting for decision making and control*, 5th ed. Boston: McGraw-Hill Irwin.

Index